MOUNTAIN WARFARE AND OTHER LOFTY PROBLEMS

Mountain Warfare and other Lofty Problems

Foreign Perspectives on High-Altitude Combat

Lester W. Grau & Charles K. Bartles
Foreign Military Studies Office • Fort Leavenworth, Kansas

Helion & Company Limited
26 Willow Road
Solihull
West Midlands
B91 1UE
England
Tel. 0121 705 3393
Fax 0121 711 4075
Email: info@helion.co.uk
Website: www.helion.co.uk
Twitter: @helionbooks
Visit our blog http://blog.helion.co.uk/

This edition published by Helion & Company 2016

Designed and typeset by Mach 3 Solutions Ltd, Bussage, Gloucestershire
Cover designed by Paul Hewitt, Battlefield Design (www.battlefield-design.co.uk)
Printed by Lightning Source Ltd, Milton Keynes, Buckinghamshire

ISBN 978-1-909982-07-9

The Foreign Military Studies Office (FMSO) at Fort Leavenworth, Kansas, is an open source research organization of the U.S. Army. Founded as the Soviet Army Studies Office in 1986, it was an innovative program that brought together military specialists and civilian academics to focus on military and security topics derived from unclassified, foreign media. The results were unclassified articles and papers that provided new understandings and broad access to information from a base of expertise in the U.S. Army, Department of Defense, and foreign and U.S. defense communities and universities. Today FMSO maintains this research tradition of special insight and highly collaborative work. FMSO conducts unclassified research of foreign perspectives of defense and security issues that are understudied or unconsidered but that are important for understanding the environments in which the U.S. military operates. FMSO's work today is still aimed at publication in unclassified journals and its research findings are taught in both military and civilian venues in the United States and around the world. FMSO is organized in the U.S. Army Training and Doctrine Command under the TRADO C G-2.

Foreign Military Studies Office 731 McClellan Avenue Fort Leavenworth, KS 66027
leav-webfmso@conus.army.mil

British Library Cataloguing-in-Publication Data.
A catalogue record for this book is available from the British Library.

The mountains are fountains of men as well as of rivers, of glaciers, of fertile soil. The great poets, philosophers, prophets, able men whose thought and deeds have moved the world, have come down from the mountains—mountain-dwellers who have grown up strong there with the forest trees in Nature's workshops.

John Muir, *John of the Mountains*, 1938

Contents

Foreword x
Editors xi
Introduction xii

MOUNTAIN COMBAT 15

1 Ground Combat at High Altitudes 17
Lieutenant Colonel (LTC) Lester W. Grau, US Army (ret.) and LTC Hernán Vázquez, Argentine Army

2 Mountain Warfare: The Need for Specialized Training 30
Major Muhammad Asim Malik, Pakistan Army

SMALL UNIT 45

3 Forbidden Cross-Border Vendetta: Spetsnaz Strike into Pakistan during the Soviet-Afghan War 47
LTC Lester W. Grau, US Army (ret.), and Ali Ahmad Jalali

4 Small Secrets of Great Mountains 56
LTC Lester W. Grau, US Army (ret.), based on the work of Ikram Nazarov

5 Flanking Detachment In The Mountains: A Soviet Experience 60
LTC Lester W. Grau, US Army (ret.) and Major V. A. Selivanov, former Soviet Army

6 Cave Warfare 67
Ikram Karimov, Mountaineer Trainer and Instructor, Uzbekistan

LARGE UNIT 73

7 The Campaign For The Caves: The Battles for Zhawar in the Soviet-Afghan War 75
LTC Lester W. Grau, US Army (ret.), and Ali Ahmad Jalali

RECONNAISSANCE 97

8 Mountain Reconnaissance Russian Style 99
LTC Lester W. Grau, US Army (ret.), based on the work of Colonel Michael Panov, Russian Army

ARTILLERY 107
9 Moving Artillery Forward: A Concept for the Fight in Afghanistan 109
Major Joseph A. Jackson, US Army

SMALL ARMS FIRE 123
10 The Right Way To Shoot: How to Select Firing Positions and Firing Ready Positions in Mountainous Conditions 125
Colonel Michael Panov, Russian Army
11 It is Difficult—But Possible and Necessary—to Master the Skills and Techniques of Shooting in Mountains 130
Colonel Michael Panov, Russian Army

SHELTER 135
12 So That Making Camp Makes You Stronger: Organizing Rest and Overnight Halts for Personnel in the Mountains 137
Colonel Michael Panov, Russian Army

RUSSIAN ARMY TRAINING 139
13 Along Slopes and Tracks: Training Method for Overcoming Mountain Obstacles 143
Colonel Michael Panov, Russian Army
14 Forward, Up, and then Down: How to Use a Rope to Climb a Steep Mountain Slope 148
Colonel Michael Panov, Russian Army
15 Don't Let Me Down—Belay: Training Method for Negotiating Mountain Obstacles 156
Colonel Michael Panov, Russian Army
16 How to "Tame" a River: Getting Across Water Obstacles in the Mountains 162
Colonel Michael Panov, Russian Army
17 Train to Win: Organizing and Conducting Physical Training For Soldiers in the Mountains 170
Colonel Michael Panov, Russian Army

LOGISTICS 175
18 Hydro-Meteorological Support of Military Transport Along Mountain Rivers 177
Vladimir Shiryayev, Military Mountaineering Instructor, Tashkent
19 What a Water Transport Company Should Be: The Need for Special Supply Delivery Transport Units in Mountain Areas 189
Vladimir Shiryayev, Military Mountaineering Instructor, Tashkent

COMMUNICATIONS 199

20 Mountain Combat: Hard to Move, Hard to Shoot, Even Harder to Communicate 201
LTC Lester W. Grau, US Army (ret.), Lieutenant Jason Falivene, USMC

MEDICAL CONSIDERATIONS 207

21 Medical Implications of High Altitude Combat 209
LTC Lester W. Grau, US Army (ret.), and William A. Jorgensen, DO MC USAR

AVIATION 219

22 Helicopter Landings in the Mountains 221
LTC Lester W. Grau, US Army (ret.), CW4 Anthony Reed, US Army, and CW3 James W. Grau, US Army, based on the work of COL D.B. Zipir, Russian Army

Conclusion 241

Further Reading 243

Index 245

Foreword

The mission of the Foreign Military Studies Office is to research and present foreign ideas and perspectives for a better understanding of present and future military and security issues. Good ideas are not ex- clusively found in the English language or even the American military experience. FMSO analysts are foreign area specialists with the language skills, military backgrounds, and the regional experience that enables them to mine foreign sources for insights that uniquely enhance our forces' understanding of the operational environment. *Mountain Warfare and Other Lofty Problems* uses Russian, Pakistani, Indian, Argentine, Afghan, and Uzbek sources.

The United States Army has been involved in a mountain war for a decade. After all this time, mountain combat remains a stubbornly difficult mission and technology can offer only modest support to the infantry's mission of closing with the enemy or the logistician's mission of getting support forward. Training, equipment, weapons, tactics, and logistics all require significant adjustment for mountain combat. This book covers operations and tactics, artillery, and aviation support, re- connaissance, communications, training, and logistics. It demonstrates that the key to enhancing effectiveness is using the terrain effectively, preparing the soldiers properly, and understanding the environment.

This book is not United States Army doctrine. Rather, it is offered as an alternate view that could help our forces successfully adapt to a most-challenging environment and carry out their mission.

Tom Wilhelm
Director, Foreign Military Studies Office
Fort Leavenworth, Kansas

Editors

Lester W. Grau is a retired infantry Lieutenant Colonel who fought in Vietnam and trained as a Soviet Foreign Area Officer. He has published over 250 articles on tactical subjects and six books on Afghanistan — *The Bear Went Over the Mountain: Soviet Combat Tactics in Afghanistan*; *The Other Side of the Mountain: Mujahideen Tactics in the Soviet-Afghan War* (coauthored with Ali A. Jalali); The Russian General Staff's *The Soviet-Afghan War: How a Superpower Fought and Lost* (cotranslated with Michael Gress); *Operation Anaconda: America's First Major Battle in Afghanistan* (coauthored with Dodge Billingsley*); Afghan Insurgent Tactics, Techniques and Procedures (TTP) Field Guide; Passing It On: Fighting the Pushtun on Afghanistan's Frontier;* and *Afghanistan: Preparing for the Bolshevik Incursion into Afghanistan and Attack on India* (cotranslated with Michael Gress and published by Helion & Company, Ltd). He has also published *The Red Army Do-It-Yourself Nazi-Bashing Guerrilla Warfare Manual (The Partisan's Companion)* and contributed to two other Helion & Company, Ltd. books *Fangs of the Lone Wolf: Chechen Tactics in the Russian-Chechen Wars 1994-2009* and *From Fabric Wings to Supersonic Fighters and Drones: A History of Military Aviation on both sides of the North-West Frontier.* Les has a doctoral degree in Military History with a concentration in Russia, Central Asia and South Asia. He is a very amateur climber with some modest peaks in the Rockies, Alps and Hindu Kush to his credit.

Charles K. Bartles is an analyst and Russian linguist at the Foreign Military Studies Office at Fort Leavenworth, Kansas. His specific research interests include: tactics, operational art, capability development, and professional military education in the post-Soviet Union. Chuck is also a captain in the Army Reserve who has deployed to Afghanistan and Iraq, and has served as a security assistance officer at U.S. embassies in Kyrgyzstan, Uzbekistan, and Kazakhstan. Chuck has a BA in Russian from the University of Nebraska-Lincoln, and a MA in Russian and Eastern European Studies from the University of Kansas, and a Geospatial Information Systems certificate from the University of Missouri-Kansas City.

Introduction

This book started as a pamphlet of recent Russian articles on mountain combat. The Russians have a lot of recent experience fighting in the mountains of Afghanistan, Chechnya, Georgia, Ossetia, and Abkhazia. Some of these are dry mountains and some are wet. This has given their force a certain edge that they do not want to lose over time. However, the Russians are not the only peoples with experience in mountain warfare, training, firing, logistics, communications, engineering, fire support, medical support, and the like. The Foreign Military Studies Office (FMSO) studies the experience of other militaries with the belief that no nation or people have a corner on good ideas. In order to provide a more comprehensive treatment and collect mountain material under the same cover, this book includes published FMSO articles on foreign mountain combat as well as recent Russian publications. It is not intended to replace US doctrine, rather it hopes to supplement US material and provide a view of a different, proven approach.

Why the emphasis on Russian experience? The Caucasus Mountains, Ural Mountains and Pamir Mountains are part of Russia and consequently, the Russian Army has a long history of mountain troops, production of specialized mountain warfare gear, and mountain training centers. Even during the Cold War, when the emphasis was on high-speed maneuver war, the Soviets still maintained mountain troops in their force structure. During the Soviet-Afghan War, the mountain warfare training requirement steadily increased until, at the end, there were seven mountain warfare training centers preparing troops prior to commitment to combat.

After the collapse of the Soviet Union, mountain warfare training — along with most other military training — slowed or stopped due to lack of funds. However, the Russian War in Chechnya demonstrated that the Russian military needed to once again promote mountain warfare training. In May 2007, *Armeisky Sbornik* (*Army Digest*) printed a series of articles on mountain warfare and mountain warfare training. We have included these articles.

The target audience of this book is the cadre of mountain warfare trainers in the United States Army and Marine Corps as well as units that are deploying to Afghanistan. There are marked differences between Russian and US mountain warfare training, The Russians put much more emphasis on weapons firing and unit movement in the mountains. They emphasize long-range logistics support using vehicles, animals, and rivers. The US Army approach is more about knots, rappelling, and individual climbing skills. The USMC trains infantry battalions at Bridgeport, California, which includes animal packing and sniper training; however, it does not train the gunners, sappers, drivers, and logisticians who are central to mountain warfare.

Mountain warfare training is not routinely part of the training a US battalion or brigade undergoes prior to deployment to Afghanistan. Consequently, most soldiers and marines lack individual and unit mountain movement skills, proper mountain equipment and the knowledge of how to use it. Soldiers and marines certainly have not faced an OPFOR in the mountains before they face a dedicated combatant in Afghanistan in a live firefight. Most soldiers and marines do not know how to fire in the mountains. Further, there are large areas of Afghanistan that cannot be reached by helicopter. A serious consequence of all this is that the Taliban has established mountain sanctuary areas inside Afghanistan, since coalition forces generally don't "do" mountains and certainly do not trek deep inside mountain ranges. This will have to change and the editors offer this collection of articles with the hope that they will help in preparing soldiers and marines for the challenge ahead.

Lester W. Grau Charles Bartles
Fort Leavenworth, Kansas

MOUNTAIN COMBAT

1

Ground Combat at High Altitudes

Lieutenant Colonel Lester W. Grau, US Army (ret.) and Lieutenant Colonel Hernán Vázquez, Argentine Army

This article first appeared in the January-February 2002 edition of *Military Review*

> *A general who allows himself to be decisively defeated in an extended mountain position deserves to be court-martialed.*
>
> Carl von Clausewitz[1]

High mountain terrain is often inaccessible, uninhabitable or of no apparent value, yet peoples and states still fight to possess it. Long, bloody wars have been fought, and are being fought, for mountain real estate located between 10,000 and 23,000 feet [3050 and 7015 meters]. Over the past fifty years, high-altitude combat has raged in Africa, Asia, and South America. The Chinese invaded Tibet in 1953 and fought a subsequent guerrilla war there until 1974. From 1953 to 1958, British troops fought Mau-Mau separatists in the Aberdares Mountains of Kenya. In 1962, China and India battled in the Himalayan Mountains bordering Bhutan and Tibet. Soviets fought Afghan Mujahideen in the towering Hindu Kush Mountains from 1979 to 1989. The Peruvian government hunted the Sendero Luminoso guerrillas in the Andes Mountains throughout the 1980s. India and Pakistan have continually battled for possession of the Siachen Glacier since April 1984 and fight sporadically over disputed Kashmir as they have since 1948. Today, Colombia's government troops are fighting the Revolutionary Armed Forces of Colombia (FARC), and the National Liberation Army (ELN) guerrillas high in the Andes, and Russian soldiers are fighting Chechen separatists high in the Caucasus Mountains.

The U.S. Army has no experience fighting in truly high mountains and its mountain warfare manuals deal primarily with low and medium mountains and stress the use of helicopter aviation to conduct that combat. However, helicopters cannot haul normal loads over 13,000 feet [3965 meters] since their rotors lack thick enough air to "bite" into, and high altitude weather conditions will frequently shut down flying for days. High-altitude combat differs from medium- and low-mountain altitude combat and requires a different orientation and force structure. Other armies have experience in truly high mountains and can provide valuable guidance and expertise. The U.S. Army needs to know how to conduct high-altitude mountain warfare, develop the tactics, techniques, and procedures to do so, and share the experience of other armies to understand and prepare for possible high-altitude conflicts.

The Environment

Mountains are generally classified as low (600 to 1500 meters), medium (from 1500 to 3500 meters) and high-altitude mountains (above 3600 meters). The world's highest mountains are not in the United States, Europe, or Korea—where the U.S. Army is accustomed to working. The Himalayan Mountain chain of Asia stretches 1,500 miles and contains nine of the world's 10 highest peaks. The Hindu Kush/Karakoram mountain chain of Asia stretches well over 500 miles with its highest peak at 28,250 feet [8,616 meters]. The South American Andes stretches over 5,000 miles and rise above 22,000 feet [6,710 meters] at many points. The Caucasus Mountains, which divide Europe and Asia, run some 700 miles with many peaks over 15,000 feet [4572 meters]. The Himalayan Mount Everest towers at 29,028 feet [8,853.5 meters] whereas the highest point in the United States, Mount McKinley in Alaska, is 20,320 feet [6,197.6 meters]. The highest point in the Colorado Rockies is Mount Elbert at 14,433 feet [4,402.1 meters]. The highest point in the European Alps is Mont Blanc at 15,771 feet [4,810.2 meters].[2]

Although high mountains occupy a good portion of the earth's surface, man is not naturally designed to live and work at these high altitudes. When a person travels to an altitude of 8,000 to 10,000 feet [2440 to 3050 meters] or higher, the atmospheric changes in pressure and available oxygen cause physiological changes, which attempt to ensure that the body gets enough oxygen.[3] These physiological changes are pronounced among mountain people who have lived in cold, high altitudes for generations. Compared to lowlanders, their bodies are short, squat, stocky, and barrel-chested, and their hands and feet are stubby. Their hearts are bigger and slower beating and their capillaries are wider. Their bodies contain 20 percent more red blood cells than lowlanders' do and these red blood cells are larger. The alveoli in their lungs are more open for oxygen absorption. Many develop a fatty epithelial pouch around the eyes to counteract cataract and snow blindness.[4] Populations at high altitude often use narcotics, such as coca or hashish, to help manage the pain and stress of high altitude.

High altitudes are characterized by extreme cold, strong winds, thin air, intense solar and ultraviolet radiation, deep snow, raging thunderstorms and blizzards, heavy fog, and rapidly changing weather, including severe storms which can cut off outside contact for a week or longer. Avalanches and rockslides are not uncommon. Although jungle or forest may hug the mountain base, trees do not grow past 10,000 to 11,500 feet [3,000 to 3,500 meters], depending on the latitude.

Physical conditions at high altitude are often more dangerous than enemy fire. Superficial bullet and shrapnel wounds can quickly turn fatal at altitude. Movement in the high mountains often results in broken bones, severe lacerations, contusions, and internal injuries caused by falls and falling rock. Frostbite and hypothermia are a constant danger. Acute mountain sickness, high altitude pulmonary edema, and cerebral edema are frequently fatal consequences of working at high altitude. Mental and physical abilities decrease at high altitude and high altitude also induces personality disorders. Sudden weight loss is often a problem. The rarefied atmosphere permits increased ultraviolet ray exposure, which creates problems with sunburn and snow blindness. High altitude shelter heating is often by unvented kerosene stoves, which means that personnel breathe air which is thick with soot.[5]

Equipment will not function, or functions marginally, at high altitudes. On the average, vehicles lose 20 to 25 percent of their rated carrying capability and use up to 75 percent more fuel.6 Military generators and vehicles are often diesel-powered, but standard diesel engines lose efficiency at 10,000 feet [3050 meters] and eventually stop functioning altogether because of insufficient oxygen. Artillery firing tables are wildly inaccurate as the changed environment allows rounds to fly much farther. Lubricants freeze; altitude and weather limit helicopters; and additional animal or gasoline-fueled overland transport adds to the physical demands and logistic requirements of this environment.

Getting There is Half the Fun

At high altitude, personnel have difficulty breathing because of decreased atmospheric pressure and subsequent rarified oxygen. Soldiers selected for high-altitude duty should be screened for their ability to function in this environment. Soldiers should be in excellent physical condition and have sound hearts and lungs. Short, wiry soldiers are preferred to tall, muscular soldiers. Selected soldiers should have above-average intelligence to allow them to more-readily adapt to the trying terrain. Personnel who have had radial keratotomy corrective eye surgery should not go to high altitudes because their vision may permanently cloud.

All personnel should undergo an acclimatization program to accustom them to their new environment and to improve their respiratory and cardiovascular systems. A physically fit soldier can adapt to the cold in about 3 weeks.[7] The body normally adapts to a higher altitude in about 2 week's time. During the acclimatization phase, the body accumulates additional red blood cells which help transport needed oxygen.[8] The Pakistani army acclimates their personnel over 7 weeks. They begin with a 3-week stay at 10,000 feet [3050 meters] where personnel acclimate to the cold while they undergo daily physical conditioning and learn mountaineering, rock climbing, rappelling, and mountain survival. During the final 4 weeks, soldiers learn advanced mountaineering techniques, trek to 14,000 feet [4270 meters] and return; trek to 17,000 feet [5185 meters] and return; and finally trek to 19,135 feet [5836 meters].[9]

Despite all training and efforts, acclimatization is not possible at heights over 18,000 feet [5418 meters], so exposure at these heights must be limited and closely supervised. Personnel at high altitudes need to be rotated out every 10 to 14 days. The Indian army acclimates its personnel over a 14-day schedule with increases in altitude at 6 days, 4 days and then another 4 days. The Indian army characteristically conducts its acclimatization by having the battalion hike from its road head to the staging area. All experienced armies agree that high-altitude acclimatization cannot be achieved in less than 10 days. An acclimated soldier is still not an experienced mountaineer. Experience counts and is not gained in 2 months of training. Some armies, such as Italy's, believe that 10 years is not too long to produce a truly capable, experienced mountain warrior.

Nothing is fast in high-altitude combat. Logistics support is key and the location of logistics dumps determines operational axes. The distance between the road head—the furthest point that supplies can be moved by truck—and the forward posts determines how many troops can actually man the forward posts. Forward posts can be a 3 to 14 day foot march from the road head. The farther the forward post is from the road head, the greater the number of troops necessary to support it. Base camps are usually built around road heads. Supplies and men travel forward from the base camps through intermittent staging posts to the forward posts. Helicopters, porters, or mules are used to move supplies from the road head. Despite attempted technology fixes, the mule is the most efficient way of moving material in the high mountains. Mules require care, attention, and training. Armies with experience in high mountains maintain trained mules and muleteers. Even mules cannot reach the higher elevations, and porters must haul the supplies forward.

Movement is calculated in time rather than distance at high altitude. Figure 1 shows average movement rates of trained, acclimated personnel and pack animals in the mountains.

The terrain slope as well as physical conditioning and altitude acclimatization of the troops determines the distance that can be covered. Figure 2 gives a rough average for determining distances over time using conditioned, acclimated troops.

Type of March	Low and Medium Mountains: *Up to 3000 meters*		High Mountains: *From 3000 to 4000 meters*		High Mountains: *From 4000 to 5000 meters*		***One Day of Rest After Marching***
	On foot	**Mounted**	**On foot**	**Mounted**	**On foot**	**Mounted**	
Normal, not affecting combat power	8	10	6	8	4	6	**6 hours**
Affecting combat power	10	12	8	10	6	8	**3 days**
Considerably affecting combat power	12	14	10	12	8	10	**1 day**
	Hours Marching Per Day						

Figure 1 Source: ROP-OO-06, *Conducción de la Brigada de Montána* (*Mountain Brigade Field Manual*) (Buenos Aires, Argentina: Government Printing Office, 1998), Annex 15, 279.

Slope Grade		Low and Medium Mountains		High Mountains		Observations
		On foot	Mounted	On foot	Mounted	
0% to 4%	Descent	4.0 Km/hr	8.0 Km/hr	4.0 Km/hr	7.0 Km/hr	Mounted troops travel walk speed
	Ascent	4.0 Km/hr	8.0 Km/hr	3.7 Km/hr	7.0 Km/hr	
5% to 9%	Descent	4.0 Km/hr	8.0 Km/hr	4.0 Km/hr	7.0 Km/hr	
	Ascent	3.7 Km/hr	7.3 Km/hr	3.0 Km/hr	6.5 Km/hr	

Figure 2 Source: Conducción de la Brigada de Montána, Annex 13, 271. Conditions for the march: day, good weather conditions, on a trail or road in good condition, firm ground, troops and pack animals trained and acclimatized.

Moving in the high mountains can be perilous. Weather can rapidly change and columns can become lost in blizzards or fog. Trail markers can quickly disappear under falling snow. Snow bridges can collapse and swallow climbers into deep crevasses. Entire patrols have disappeared without a trace while moving to the Siachen Glacier.

Line-of-sight communications is excellent in the mountains but difficult to achieve because of high peaks. Therefore, communications sites are carefully selected and often become key terrain. Very-high frequency radios with automatic frequency hopping, encryption, and burst transmission capabilities work best. Normal batteries quickly lose power in the cold, so lithium batteries should be the normal issue.[10] Frequently, mountain tops become part of the national communications infrastructure because they are crowded with military, national, and commercial radio and television sites and telephone relay towers. These vital areas need to be protected, and military platoons often garrison such communications sites against guerrilla attacks.

Combat at Altitude

There are two primary scenarios for combat at altitude. First, two states dispute the boundary between their countries and maintain forces supporting a rough line of demarcation along the disputed zone (Kashmir and Siachen Glacier between India and Pakistan, and the Kameng Frontier Division between India and China). In this scenario, opposing forces hold linear defenses along the line of demarcation, regardless of altitude, and conduct a fairly positional fight.

Second, a light infantry force of guerrillas, smugglers, bandits, or forces from a neighboring state transverses the mountainous region to establish base camps in the mountains from which they patrol, launch raids, or maintain smuggling routes. This has been the pattern for a number of high-altitude disputes such as the Mau Mau uprising, Soviet-Afghan war, Sendero Luminoso in Peru, Russo-Chechen wars, and Colombian efforts against the FARC and ELN. In this scenario, the fighting does not automatically gravitate to a border zone, but usually stays below the tree line.

At altitude, the first enemy is the environment. The second enemy is the human foe. At altitude, high ground is not always key terrain. Frequently, key terrain is related to mobility—passes, main supply routes, road heads, and intermittent staging posts. Light infantry and artillery are the primary combat forces.

Offensive actions in the mountains include infiltration, ambush, raids, patrolling, shelling attacks, limited air assault, and limited offensives. Pursuit is seldom possible. Envelopment is the most common maneuver and the frontal attack is the least desired option. Defensive actions include counter-infiltration, ambush, patrolling, and positional defense. Relief in place is routine small-unit action.

Offensive actions should focus on interdicting logistics by blocking passes, denying use of supply and transit routes, capturing base camps and intermittent staging posts, and destroying transport.[11] Force oriented offensive actions, such as interdicting patrols or raiding artillery positions, make great headlines and can boost morale, but they seldom have the long-term effect as actions against logistics. Offensive actions are small-unit actions, since only small units can be supported at altitude, and frequently the terrain is so restricted that too many soldiers would hamper the effort. Movement is by small groups moving at a walk to avoid sweating because sweat freezes quickly leading to frostbite. Objectives are close at hand so the attackers will not be exhausted before they arrive and will not be caught in the open by rapidly changing weather. Assembly areas may be nonexistent and the attackers will have to move directly from forward positions. The attack may have to go in waves if suppressive fire is inadequate or the enemy is conducting a reverse-slope defense. The offensive plan must be clear, as most mountain maps are problematic. Maneuver is dictated by terrain and the reserve is committed early since movement is slow and mutual support is very difficult to achieve. Maneuver is slow and limited in distance. A maneuver force can range from one or two men to a full battalion if weather and the enemy situation permits.

Defending at altitude is difficult because of limited troops and material. When defending along a border, a battalion holds an extended frontage (7,000 to 8,000 yards) while a company holds 1,500 yards, so there is little depth, or large gaps, in the defense.[12] Further, the complete battalion is seldom on line simultaneously. Often, a platoon holds a company position since the rest of the company is being held in reserve at lower elevations where the deterioration of the body is not as rapid. The platoon is rotated every 10 to 14 days. The entire company must still be rotated to lower elevations to recoup every 3 to 4 months. This means that the long, linear defense is actually a string of strong points built around a machine gun. Reverse slope defense, with forward slope observation posts is preferred, since the defensive positions often lack overhead cover and are susceptible to artillery airburst.

A great deal of daily effort is required to keep snow from completely filling the defensive positions and hiding the trails. Permanent shelter, such as portable fiberglass huts, are essential at the defensive positions.[13] Fortifying defensive positions is difficult since this usually requires the delivery of heavy materials such as cement, sand, water, and roofing timbers. Sensors are a welcome addition to the defense in those areas where they will not be rapidly covered by snow. Defensive positions should be designed and stocked to hold out independently for days since relief in the mountains is problematic due to weather. Conversely, when the enemy is a guerrilla force, the defensive position is a perimeter defense from which patrols, ambushes, and raids are launched.

Mountain patrolling is a common feature of the offense and defense. Small patrols are at risk, so platoon-sized patrols are common. Single patrols are useless, so multiple patrols are normal. Local guides or scouts are an essential part of each patrol. Detailed planning is an essential part of the patrol plan and includes a reaction force or reserve. The meeting battle is normal combat at altitude resulting from probing actions by opposing patrols.

Raids are a common offensive and defensive tactic. They are designed to seize a point, exploit success, and then withdraw. Raids are a temporary measure to capture personnel and equipment, destroy installations, bait traps to draw enemy reaction, and attack morale. Since there is no intention of holding the objective for a length of time, the logistics burden is less onerous than a deliberate attack. Successful mountain raids normally incorporate an assault force, a fire support group, and a security element.[14]

Fire Support at Altitude

Mountains restrict effective bombing and strafing by jet aircraft. It is difficult for them to pick out targets that are camouflaged or concealed by natural cover. Weather, deep shadows, and the environment also restrict pilots' vision. There are few approach routes and most of those are along valleys, which are covered by air defense and infantry forces using massed fire. Climate and terrain restrict jet aircraft from diving freely or flying low enough to engage targets effectively. Still, camouflage discipline, controlled movement, and layered air defense are essential to prevent savaging by high-performance aircraft.[15] Helicopter gunships are more of a danger to ground forces, but eventually altitude limits their effectiveness. Lightweight helicopters can serve effectively as artillery spotters. All aviation is subject to the vagaries of weather at high elevation, which is powerful, constantly changing, and often shuts down flying. Dense fog, high winds, and blizzards are common and whiteouts are a constant threat to pilots.

Artillery remains the round-the-clock fire support system. However, artillery is often constrained during high-altitude combat. Sharp bends, high gradients, and the general condition of mountain roads restrict the movement of artillery, towed guns in particular. There are a limited number of gun positions, so artillery batteries are seldom deployed intact. One- and two-gun or rocket launcher positions are common. Consequently, the number of alternate firing positions is also restricted and these positions tend to become permanent. Guns should be moved at night for protection against enemy aircraft and artillery. However, night movement of guns in mountainous terrain is risky and accident-prone. Artillery positions should be constructed so that gun crews can defend them against ground attack. Firing positions should be on reverse slopes and as close to the crest as possible—considering crest clearance and flash-cover. Individual guns should be sited in terrain folds and other places where they are naturally concealed.[16] Artillery plays a major role in logistics interdiction, counterbattery and shelling front-line units. Artillery can create havoc with a forward defense by targeting living accommodations and using airbursts against troops in the open.[17] Mortars are frequently more effective than guns or howitzers. They are easier to shift around, can better engage reverse slopes and can be moved closer to the forward posts.

Transport frequently determines the location of artillery and mortars and the supporting range of artillery. Artillery cannot be readily moved where there are not roads. Artillery firing points are usually located where ammunition can be delivered—in valleys, villages, and near road heads.

Logistics Support

High-altitude logistics are key since the terrain and unique environment hamper delivery to the forward troops. Logistics always drives the battle, but in high-altitude combat, this is especially so. Without good highways or railroads, dump sites cannot be readily moved, it takes an inordinate amount of time to shift troops from one sector to another, and logistics demands are considerably higher than in other types of light infantry combat. Trucks, helicopters, mechanical mules, and snowmobiles are key to mountain logistics, but above 13,000 feet, the logistics effort shifts to the backs of mules and porters. Naturally, this is the point where the logistics delivery system snarls since porters and mules have distinct limitations and there are never enough of them.

Trucks are important to logistics support and gasoline-powered trucks are clearly preferred over diesel. As the truck ascends the mountain, the amount of oxygen available is reduced and the engine efficiency drops off. Cross-country and climbing capability decline as fuel usage soars. Diesel engines may need to be fitted with turbo-chargers and gasoline engines may need their carburetors adjusted. Figure 3 shows the average increase in fuel consumption at altitude.

Helicopter-based logistics are the preferred mode in mountain warfare, but the mountains are not the optimum helicopter environment. Air density decreases with altitude and mountain winds and updrafts are unpredictable and dangerous. Proper

			Light Vehicles (Gasoline)	Trucks 3 to 5 Tons (Diesel)		Observations
		Type of Slope	Average Load	Loaded	Unloaded	
Low Mountains		3 to 5%	16	25	20	Good engine performance
		6 to 8%	17	30	25	
Medium Mountains	Up to 3000 meters	3 to 5%	17	30	25	
		6 to 8%	18	32	27	With good roads
High Mountains	3000 to 4000 meters	3 to 5%	18	32	27	Head winds increase consumption by 10%
		6 to 8%	20	35	30	
	Above 4000 meters	3 to 5%	20	35	30	
		6 to 8%	25	38	33	

Figure 3 Source: Conducción de la Brigada de Montána, Annex 16, 281.

landing zones are difficult to find and, if close to the enemy, probably under enemy mortar and small-arms coverage. Helicopters must follow the terrain features of the mountains adding predictability to their approaches and increasing the risk to the crew. Fog, sudden storms, icing, and variable winds can quickly shut down helicopter support. Mountain terrain interferes with air-toground communications and with air-to-air communications. Planning for helicopter support in the mountains requires detailed planning, first-rate liaison, and a habitual association between the helicopter and ground unit encompassing training and social events. Flying in the mountains is so different that the armies of India, Pakistan, Columbia, Argentina, and Switzerland have special mountain flight courses for their helicopter crews.

A Step Back in Time

High mountains are counter-technology. Mules are a good option for high-altitude logistics. They can use very narrow trails, can carry more than a human porter, and tire less over long distance. American mules can carry up to 20 percent of their body weight (150-300 pounds) for 15 to 20 miles per day in mountains.[18] Smaller mules in other locales will carry less. The maximum carrying weight for an Argentine mule is between 200 and 250 pounds. However, this is for low- and medium-altitude mountains. At high altitude, the maximum carrying weight drops below 200 pounds. Organized mule cargo units, rather than ad hoc teams led by local teamsters, are the preferred option, but local mules are always preferred over deployed mules.

Mules were part of the U.S. Army during World War II in Burma and Italy and were a critical element of the Mujahideen supply effort in the Soviet-Afghan war. They remain part of the force structure of many contemporary forces with high-altitude mountain troops. Other armies contract mule transport through local teamsters. Yet mules have their limitations. If the snow is too deep, they simply refuse to move.

American mules require 10 pounds of grain and 14 pounds of hay per day, which also becomes part of the logistics load.[19] The smaller mules of Argentina require eight

Rations	1,620 kilos	Assault & field rations
Ammunition	1,021 kilos	One basic load-circumstances may require carrying more
Water (drinking and cooking)	2,262 kilos	2,565 liters
Total weight (for company)	4,903 kilos	Not counting food, water, ammo, and packs carried by soldiers
Mules (to haul weight)	61 mules	Small mules haul 80 kilos; Big mules up to 150 kilos
Mules (to haul crew-served weapons)	20 mules	Machine guns and mortars
Mules (to haul fodder, grain, salt and water for mules)	151 mules	5,620 kilos of fodder, 4,060 kilos of grain, 348 kilos of salt, and 18,416 kilos of water

Figure 4 Source: Major Valero, Army of Argentina training exercise material.

pounds of grain and eight pounds of hay per day. Mules consume 25 to 30 liters of water a day and up to 50 liters in desert country. They also require a daily ounce of salt. Like humans, mules require time to acclimate to altitude. Muleteers and mules require about a month's training to get them ready to work above 3,000 meters. Like humans, mules tire easily above 4,000 meters and need to be rested frequently. Mules also have to be trained not to fear the noise of firearms and explosives so that they do not run off during a march.[20]

Mules are subject to colic, heat exhaustion, injuries, and wounds. Most injuries and wounds result from poorly adjusted saddles, pack frames and harnesses. Stones, rocks, and debris on the trail can also wound a mule's hoof. Local mules are more immune to disease at altitude than humans and all mules have a keen sense of self-preservation that keeps them alive in mountain storms.[21] Mules require a great deal of daily care and training. Muleteers, farriers, blacksmiths, and large animal veterinarians, who have been absent from many armies for decades, are essential for mule-borne logistics. Mules need new shoes every 30 days and there are special mule shoes for snow and ice. Figure 4 shows the supply and transport estimate for a 171-man light infantry company planning a mountain march, attack and defense lasting for a total of 6 days. Since much of the material will be kept in dumps and moved in stages, the commander has managed to keep his transport requirements in hand.

Porters should be hired from the local populace since they are acclimated to the elevation and are accustomed to moving around the mountains safely. Locals used to carrying loads have developed endurance and are accustomed to breathing thin air. Although a porter cannot carry as much as a mule, they can move in places where mules cannot. However, porters will probably be reluctant to work too far away from their homes and villages. There is always a security consideration when using local porters. Figure 5 shows porter-carrying capabilities.

During the Peru-Ecuador border conflict for the Condor Cordillera in 1994, the Peruvian army relied on porters exclusively for resupply.

Although the fight was in medium-altitude mountains, not over 2500 meters, the forward logistics support was restricted to porters because the steep mountains were covered with thick jungle, had few trails, and the Peruvian army lacked trained mules and muleteers. The Peruvian army moved its supplies from one small village to the

	Water				Wood		
	Man		Animal		Cooking	Heating	
	Drinking	*Cooking*	*Summer*	*Winter*	*Uncovered*	*Tent/shelter*	*Open*
Low and Medium Mountains	1.5 to 2 Liters	5 Liters	15 Liters	10 Liters	1 kg per man	1 kg per man	1 kg per man
High Mountains	2 to 2.5 Liters	8 Liters	15 Liters	10 Liters	1 kg per man	Every 6 hours	Every 2 hours for fire

Figure 5 Source: Conducción de la Brigada de Montána, Annex 17.

next, using local villagers as porters to carry the supplies eventually to the fighting up on the Condor Cordillera.

Front-line combatants need daily supplies of ammunition, food, water, and heat for survival. Figure 6 shows daily consumption rates of water and wood fuel.

In the mountains, a battalion task orce tries to carry and stockpile enough supplies to operate for 1 to 2 weeks. This requires expending time and energy to establish supply dumps along the main supply route. Naturally, the shorter the supply route, the easier it is to protect. If roads, tracks, and trails are under enemy control, the unit might be restricted to helicopter supply and its inherent problems in the mountains. Logistics support at higher altitudes during winter may become impossible causing opposing forces to withdraw.

Medical evacuation at altitude is frequently difficult. Weather or weight limitations may prevent a helicopter from flying to a patient. Often, patients must be carried on stretchers to lower elevations where the helicopters can reach. Soviet experience in the mountains of Afghanistan proved that 13 to 15 men might be involved in carrying one patient. Exertion at altitude is difficult and the stretcher party has to provide its own security as well.[22] Patients cannot be effectively treated at altitude, but have to be evacuated to lower altitudes to survive.[23]

The Eternal Mountains

Mountain terrain is difficult, movement is slow and the hazards to health and physical well being are significant and constant. Combat at high altitude is a historical constant and a contemporary fact. It cannot always be avoided. Training for mountain combat is not simply light infantry training. Special training and acclimatization is necessary.

Leadership is particularly important in mountain combat. The harsh living conditions, physical deterioration, and psychological depression inherent in mountain combat require skilled leaders. Armies with regimental systems and years-long association find it easier to cope with the leadership challenges of mountain combat. Combat is primarily small unit, placing a great deal of responsibility on platoon and squad leaders.

Fire support is difficult. Artillery firing tables are inaccurate and artillery is hard to move on mountain roads. Transporting guns by helicopter is recommended where

	Water				Wood		
	Man		Animal		Cooking	Heating	
	Drinking	*Cooking*	*Summer*	*Winter*	*Uncovered*	*Tent/shelter*	*Open*
Low and Medium Mountains	1.5 to 2 Liters	5 Liters	15 Liters	10 Liters	1 kg per man	1 kg per man	1 kg per man
High Mountains	2 to 2.5 Liters	8 Liters	15 Liters	10 Liters	1 kg per man	Every 6 hours	Every 2 hours for fire

Figure 6 Source: Conducción de la Brigada de Montána, Annex 18.

possible. Moving guns and ammunition takes an unusual amount of time. Helicopter gunships provide excellent support at lower altitudes. Mortars are excellent for hitting reverse slope positions, but have limited range.

Logistics are a primary concern in mountain combat with transport to altitude requiring special effort. Sustained combat requires an inordinate logistics effort. Small-unit actions, where units do not remain for extended periods of time, do not impose the same logistics burden.

Although the U.S. Army has not fought at truly high altitude, this may not always be the case. High mountains occupy much of the world's surface and they are not immune to the world's conflicts. Other nations have successfully fought at altitudes above 10,000 feet. Should the U.S. Army find itself committed at these altitudes, the experiences of other nations are invaluable. Preparation for such an eventuality should begin well before crisis dictates deployment.

At the time of original publication, Lieutenant Colonel Herán Vázquez was the Argentine Army's liaison office to the US Army Combined Arms Center and consulting editor for the Hispano-American edition of Military Review, *Fort Leavenworth, Kansas. He is a graduate of the Argentine Army War College. He has served in a variety of infantry and mountain units in Argentina. He served as S3 of a Motorized Light Infantry Battalion in the Patagonia Desert; S3 of the Mountain School Battalion; two tours as a member of the United Nations Peacekeeping Force in Cyprus with the Joint Operations Center, Operations Department, Headquarters Peacekeeping Force; operations officer for Argentine Task Force 10 "Chipre"; and tactics and MDMP instructor at the Argentine Army War School. He is a qualified mountain troop leader, commando and paratrooper.*

Notes

1. Carl von Clausewitz, *On War*, ed. and trans. by Michael Howard and Peter Paret (Princeton, NJ: Princeton University Press, 1989), 432. The authors thank mountain warriors Lieutenant Colonel (LTC) German Giraldo, Army of Colombia; LTC Foto Duro, Albanian Army; Major (MAJ) Alejandro Valero, Army of Argentina, MAJ Akbar Khan, Army of Pakistan; and LTC Tejbir Singh, Army of India, for their input, suggestions, critique, and guidance. The authors retain responsibility for the accuracy and ideas in the article.
2. *The Houghton-Mifflin Dictionary of Geography: Places and Peoples of the World* (Boston, MA: Houghton-Mifflin Company, 1997), 48, 164, 165, and 450.
3. Tim J. Setnicka, *Wilderness Search and Rescue* (Boston, MA: Appalachian Mountain Club, 1980), 620.
4. Syed Ishfaq Ali, *Fangs of Ice: The Story of Siachen* (Rawalpindi, Pakistan: American Commercial PVT Ltd., 1991), 15-16 and 110-11.
5. Lester W. Grau and William A. Jorgensen, "Medical Implications of High-Altitude Combat," *U.S. Army Medical Department Journal*, to be published April 2002.

6. Aleksey Svetlanov, "Osobennosti razvedki v gorakh" (Peculiarities of Mountain Reconnaissance), Voyennye znaniya (*Military Bannerr*) (June 2000), 13.
7. Salman Beg Punjab, "Operations in Glaciated Areas," *Pakistan Army Journal* (Spring 1994), 5.
8. Setnicka, 623.
9. Ali, 16-17; Punjab, 5-6.
10. Punjab, 7.
11. Ibid., 8-9.
12. Iftikhar ur Rahman, "Defensive Battle in Mountains," *Pakistan Army Journal* (September 1984), 31.
13. Punjab, 4.
14. Ali Ahmad Jalali and Lester W. Grau, *The Other Side of the Mountain: Mujahideen Tactics in the Soviet-Afghan War* (Quantico, VA: U.S. Marine Corps Study: DM: 980701, 1998), 69-103 and 211-25.
15. Sikandar Hameed, "Local Defence of Gun Area in Mountainous Terrain," *Pakistan Army Journal* (March 1983), 35-41.
16. Ibid.
17. Punjab, 10.
18. U.S. Army Field Manual 31-27, *Pack Animals in Support of Army Special Operations Forces* (Washington, DC: U.S. Government Printing Office, February 2000), 1-2.
10. Ibid., 2-14.
20. Reglamento Funcional Público (Public Functional Regulation) 24-02, Reglamento Ganado de Servicio (Service Livestock Regulation) (Buenos Aires, Argentina, 1994), 36.
21. Ibid, Annex 18.
22. Boris Gromov, *Ogranichennyy kontingent* (Limited Contingent) (Moscow: Progress Publishers, 1994), 186.
23. Grau and Jorgensen.

2

Mountain Warfare: The Need for Specialized Training

Major Muhammad Asim Malik, Pakistan Army

This article first appeared in the September-October 2004 edition of *Military Review*

Mountain Warfare is specialized combat with unique characteristics. Military leaders and soldiers need training and experience to understand the peculiarities of mountainous environments and how they affect combat. Armies that train for mountain combat perform much better than those that do not.

During World War II, the German Army raised an entire corps of elite mountain troops called *gebirgs jaeger* (mountain troops). Although not all of these troops were used in the mountains, they demonstrated superior abilities in almost all theaters in which they were used. The German Fifth Mountain Division marched more than 248 miles, crossed mountain passes above 6,500 feet, and secured well-entrenched defenses on the Mestksas Line.[1] Other *gebirgs jaeger* captured most of the Caucasus mountain region in the summer of 1942.

The US Army 10th Mountain Division, trained in the mountains of Colorado during World War II, captured Riva Ridge in northern Italy. US Army Field Manual (FM) 3-97.6, *Mountain Operations*, cites this action as clearly demonstrating the value of superior skills in mountain warfare.[2] The importance of preparing leaders and soldiers for high-altitude combat cannot be overemphasized.

The Pakistan Army has been in Kashmir for more than 50 years and has a great deal of experience in fighting in mountainous, glaciated terrain. While the US Army has quality training institutions for mountain warfare, it does not have many troops with extensive, current mountain-combat experience. A comparison of the training techniques of US and Pakistan armies provides insight into preparing for mountain combat.

Training Prerequisites for Mountain Warfare

Physical fitness is the first prerequisite of mountain-warfare training. The effects of cold weather and unforgiving terrain require a high level of physical fitness for long-distance climbing and walking, and the physical fitness required for mountainous terrain must be developed at high altitude. But being physically fit does not necessarily mean soldiers will be able to perform adequately at high elevations. US soldiers selected to attend the mountain-warfare school in Kakul, Pakistan, required additional climbing time to attain the desired level of physical fitness. The body must adjust to the thin mountain air, and climbing muscles must be developed.[3]

Mountainous terrain can be an ally or a dangerous adversary. In Kashmir each year, thousands of troops are introduced to the mountainous environment to help them understand and appreciate it. A marked difference exists in the performance of units that have conducted vigorous acclimatization training and those that have not. Weather and terrain-related casualties are a big indicator. During initial training in Colorado during the early 1940s, the US 10th Mountain Division suffered more casualties from weather-related injuries than from actual mountain combat in Italy.[4]

Field Manual 3-97.6 defines high mountains as those above 3,000 feet; however, it does not fully explain the high-altitude environment. Despite the fact the Pakistan Army has been fighting in Kashmir for a long time, operations at higher altitudes (18,000 to 22,000 feet) were not fully understood before the Siachen conflict between India and Pakistan in 1984.[5] During the conflict, frostbite, sunburn, and other high-altitude sicknesses caused large numbers of casualties. Such incidences are now rare, however, because troops and commanders are trained to take precautionary measures.

In Kashmir, stone or wooden bunkers, which double as living accommodations and fighting bunkers, are found at posts below 13,000 feet, but at high altitudes, stone structures are not practical. Cement will not bind, and the underlying glacier is always moving. Instead, prefabricated, synthetic domes (igloos) are used.[6] The domes are easy to carry and assemble even at 18,000 feet and above. They can be retrieved from even large amounts of snow and set up again quickly.

Soldiers must also be trained to wear proper clothing. Loose-fitting layers and insulated and polypropylene clothing that does not allow perspiration to accumulate close to the body are best. Developing frostbite from touching metal equipment with one's bare hands is possible when temperatures drop to minus 22 degrees Fahrenheit. Pressurized sleeping bags help stabilize soldiers suffering from altitude-related sicknesses.[7]

Basic mountaineering and high-altitude skills are vital for soldiers to develop confidence and survive in mountainous environments and essential in combat. As mountain height increases, so does the required skill level. At altitudes below 13,000 feet, it might be enough for soldiers to understand climbing techniques, navigation, route selection, the use of ropes, and procedures to avoid landslides and snow avalanches, but at high altitudes, soldiers must learn more complex techniques, such as those required for mountain expeditions.[8]

Because it is not always possible to transport material by helicopter, troops are often required to carry awkward loads, including kerosene oilcans, rations, and building materials for bunkers. The Soviets learned this lesson while fighting in difficult terrain in Afghanistan.[9]

At high altitudes, where it is difficult to keep weapons functioning, covering and protecting weapons and equipment against snow and ice is a necessity. Batteries often will not perform optimally in the cold, and complicated mechanisms, such as those in surface-to-air missiles, can easily malfunction. Also, artillery shells sometimes behave erratically because of thin air and gusting winds.[10]

Surviving and operating in mountainous terrain requires more energy than usual. A soldier who needs 3,000 to 4,000 calories under normal circumstances will require 6,000 or more calories in the mountains. To complicate the situation, high altitude adversely affects a person's appetite. Soldiers tend to eat and drink less in high altitudes, which reduces morale and fighting capabilities and makes them more susceptible to mountain-related illnesses.[11] US soldiers conducting mountain-warfare training at Abbotabad, Pakistan, which is at 4,000 feet, lost approximately 25 pounds during a 3-week training period.[12] Commanders must ensure soldiers consume proper diets and are well-hydrated.

Physiological and psychological effects become more pronounced at altitudes above 8,000 feet. Soldiers must take preventive measures and be trained to detect signs of illness in colleagues. Common symptoms include severe, persistent headaches; coughing; difficulty in breathing; and aloofness. Other symptoms might include swelling around the eyelids, incoherent speech, intolerance, and even outright aggressiveness.[13]

Many problems, including altitude sickness, can occur during the initial stages of a soldier's arrival at high altitude. The biggest killer, cerebral or pulmonary edema, is difficult to detect, but often develops if soldiers stay too long at high altitude.[14]

The normal practice in glaciated areas is to not keep soldiers above 19,000 feet for more than three to four weeks before returning them to lower elevations. If soldiers experience any signs of altitude illnesses, commanders must evacuate them promptly. For most mountain illnesses, evacuation to below at least 3,000 feet is the first requirement for saving a person's life.[15] Delaying evacuation might not only cost the soldier's life, but imperil the lives of the soldiers who might have to conduct evacuation procedures during bad weather.

Replacements being sent to high-altitude environments must have operated at heights similar to those to which they are being sent for at least 10 to 15 days. If not, they could quickly become casualties themselves. Well-trained, acclimatized troops must be available to replace those at higher altitudes.

High-altitude environments can take heavy physical and mental tolls on soldiers. While in the Caucasus, the Germans learned that troops wore down much faster in mountains despite the fact they were elite troops, picked for their mental abilities and physical prowess. Operations in such environments involve extreme physical exertion.[16] Living conditions in mountainous terrain can be difficult. At times all movement is stopped, soldiers do not receive mail, and replacements might not arrive on time. These factors can lead to depression and boredom and a sharp decrease in fighting spirit. Simple tasks such as manning weapons, sentry duty, and patrolling require determination.

Offensive actions in mountainous terrain are difficult and costly. Not only must soldiers fight the enemy, they must also brave the elements of harsh terrain, which are equally formidable. These conditions call for strong leadership by junior leaders, who must physically lead and be mentally tough.

Leadership

Mountainous and high-altitude environments are extremely demanding and require a high level of leadership at the small-unit level as well as at higher levels. Leaders must understand the constraints placed on their soldiers' performances and should include terrain and weather in planning and executing actions to avoid miscalculating the timeframe, logistic requirements, and force capability.

Mountain combat is decentralized and often takes place at the platoon or squad level. The quality of junior leadership is decisive. The Russians observed in Afghanistan that even a small unit, maneuvering boldly, could decide the outcome of a battle.[17]

A hallmark of German alpine troops during World Wars I and II was the quality of their leadership. German Field Marshal Erwin Rommel, who began his career in a German mountain regiment during World War I, relentlessly advanced his small detachment in difficult terrain capturing one hilltop after another.

Mountainous environments demand that junior leaders set the example of physical fitness and endurance. During the Pakistan-India conflict at Kargill in northern Kashmir in 1998, Captain Sher Khan was posthumously awarded the "Nishan-E-Haider," Pakistan's highest military award, for conducting daring raids with a handful of men against enemy patrols and convoys. His last action included a successful counterattack to recapture a post against overwhelming odds. Such leadership inspires subordinates and raises the morale of the entire unit.

At high altitudes, small-unit leaders must be trained to recognize and address physical and psychological fatigue in their soldiers, including loneliness, depression, and violent mood swings. The relationship between officers and troops is more intimate than under normal circumstances, so leaders should be able to identify signs of deterioration in soldiers before they become pronounced.

Decisions at higher levels influence the conduct of tactical actions. The German Army emphasized the importance of meticulous planning and preparation even for small-scale operations.[18]

Field Manual 3-97.6 emphasizes realistic timetables based on reconnaissance and the commander's practical knowledge of the mountain battlefield.[19] Planners must understand that mountainous terrain adversely affects time and space calculations—an important lesson learned during the US Marine Corps (USMC) exercise Alpine Warrior at Fort McCoy, Wisconsin, in 1986.[20]

Cost-effective mountain combat requires skilled and well-trained troops. Soldiers cannot be sent into a fight at high altitude at the last moment. Doing so could invite disaster. One example of such an action is the employment of the 7th Indian Brigade against the Chinese in the 1962 Himalayan conflict.[21] The brigade had not been stationed in the mountains previously, and when things began going badly, the brigade was moved from the plains straight into mountain combat. The soldiers, who had not been acclimatized or equipped to fight in the mountains, suffered heavy casualties because of frostbite, edema, and other highaltitude-induced illnesses.

Communications

Terrain and unpredictable weather conditions affect communications at high altitudes. Satellite communications and the use of command and control (C2) aircraft can offset some terrain limitations and reduce reliance on bulky radio equipment. Crews responsible for installing and maintaining retransmission stations— often situated on the highest peaks to provide adequate range and coverage must be well trained in mountain-survival techniques. Also, these isolated stations are targets for guerrilla bands, as the Soviets discovered when they tried to protect similar sites in Afghanistan.[22]

Fire support

Mountainous terrain significantly influences artillery fire support. Targets are located on peaks, in ravines, and on reverse slopes; no continuous front exists; and weather conditions are unpredictable. Undulating terrain and intervening crests require a large number of observers located on dominating heights to cover the entire area of operations. Gun positions that are ideal for range and coverage might not be suitable because of intervening features and masking fire. At other times a location might be tactically sound but will be an area prone to avalanches or flash floods. Once guns are deployed, major engineering and logistical efforts might be necessary to shift them to alternate locations in a timely fashion.

Air burst and variable time fuzes are more effective than point-detonating artillery rounds. Howitzers and mortars are more effective because of their ability to engage targets on reverse slopes. The Taliban used mortars to hit US troops successfully during operations in Afghanistan.[23]

Maneuver

Mountainous terrain is ideally suited for the defense. During World War II, some of the heaviest casualties in the Italian Theater occurred during an attempt to overcome German defenses at Mount Casino. In Afghanistan, the Russians attacked the strategic Panjshir Valley repeatedly but were unable to clear it despite their advantage in firepower and mobility.[24] The line of control in Kashmir in 2003 was not much different from the cease-fire line of the India-Pakistan war in 1949.[25] Both Indian and Pakistan forces found that an assault on well-defended positions was extremely costly. Defense requires the control of dominating heights, passes, and lines of communication by strong-points. An integrated defense is not possible in cut up, mountainous terrain. During training, commanders need to understand the techniques of defense with all-around protection and emplacement of direct fighting weapons. Field Manual 3-97.6 highlights that reserves must be closer to important defense locations because reaction times in mountainous terrains are longer than usual, which could require several small rather than one large centralized reserve.[26]

Mountainous terrain offers opportunities for infiltration, requiring defenders to be aggressive at all times. Aggressive patrolling enhances security and keeps soldiers active and sharp. In Kashmir this helped prevent a bunker mentality. Although sensors provide some protection, mountainous terrain is too compartmentalized for complete electronic surveillance. Combat service support (CSS) elements must provide their own protection and must train in patrolling and perimeter defense while developing a mindset focused on constant vigilance.

Offensive operations require meticulous planning and preparation because of the inherent strength mountainous terrain provides to the defender. Training plays a vital role in ensuring an edge for the attackers. Since the defender has an advantage, successful attacks should isolate the defender and keep him under constant pressure. The Soviets laid great emphasis on junior leaders and company-level mountain operations, advocating envelopment by smaller, autonomous groups.

During Operation Anaconda in Afghanistan, US forces used more decentralized combat than on normal terrain. Junior leaders' initiative and skill is vital to the mission's success, especially in security and reconnaissance missions. Mountainous terrain and bad weather provide opportunities for small forces to concentrate and achieve surprise. Russian and Afghan government forces suffered heavily when they neglected this aspect of the battlefield environment.[27]

The Soviets used helicopters in Afghanistan to airlift troops and supplies into battle.[28] Helicopter gunships effectively supported ground operations until the Mujahideen obtained Stinger missiles, which tilted the situation in their favor.

US forces also rely on helicopters for transportation and movement in the mountains, requiring aviation planners to be involved in the planning process early. With beyond-line-of-sight and precision-guided munitions, aviation and air assets have neutralized many inherent problems in mountain warfare, but they have not eliminated the need for specialized training. Because mountain combat tends to be decentralized, control of supporting fire is more difficult. Tight control of jet aircraft and helicopter gunships is necessary to avoid fratricide.

Logistics

Logistics support in the mountains is difficult and time-consuming. In Kashmir, a variety of transport is used for logistical support, road transport being the most reliable and cost-effective. At higher altitudes where tracks cannot be maintained because of snow and difficult terrain, mules are a preferred means of transport.[29] At altitudes where even mules cannot go, porters can. Porters are local people capable of carrying heavy loads across difficult terrain.

In the Caucasus Campaign, the German army used sleds, mules, and horses in addition to trucks.[30] Recently, despite technological advances, the US Army had to use horses and mules in Afghanistan. Helicopters are a quick, versatile means of transportation, but at higher altitudes their lift capability is severely limited. The French Alouette helicopter can fly higher than U.S helicopters can, but even it can deliver only about 180 pounds above 20,000 feet. Because helicopters cannot be used in adverse weather, a mixture of resources is necessary to ensure reliability and flexibility.[31]

The road network in the mountains is generally a logistician's nightmare. Main supply routes are limited and often do not support vehicles that require large turning radii. Many roads do not permit two-way traffic.

While tactical plans take into account main roads, tactical engagements do not usually occur close to roadheads. At Siachen, the Pakistan Army built roads near forward defenses, but the real challenge was in transporting supplies across the last few miles from roadheads to forward posts.

The Center for Army Lessons Learned (CALL) analysis of the operation in Afghanistan recognized the need to have logistics as far forward as possible: "It might require additional staff work from the logisticians to deploy the logistics to the work area (like rations to the platoons, mortar rounds to the mortars), but the advantage is reduced expenditure of energy for those on the ground."[32]

Logistics estimates and loads must be customized for the mountainous environment. For example, using mules requires loads be broken up according to their carrying capacity. Also, overages must be built into supply estimates because there is always a need for a large reserve of items that wear out quickly, such as boots, jackets, and gloves. If soldiers use improper or worn clothing for even a short time, the chance of developing altitude and cold-related sicknesses increases significantly. In addition, combat casualty evacuation involves many challenges. Air evacuation remains the preferred method, but because of the dispersed nature of troops, expert medical help might not be available quickly. Therefore, self-aid, buddy help, and the availability of more combat life savers in the unit is important.[33]

Canadian small-unit support vehicles, specially designed for restrictive terrain, were particularly useful for logistics support at high altitude in Afghanistan, whereas the bulky ground-held laser designating system was not. Soldiers' personal loads of more than 50 pounds were too heavy at high altitudes. Equipment must be upgraded for future mountain warfare.[34]

Training for Mountain Warfare

Recent operations by US and other Allied forces confirm the need for specialized mountain-warfare training. CALL's analysis recognizes that soldiers with mountain experience exhibit exceptional morale, physical stamina, and technical competence in decisive combat operations. The analysis also recognizes that coalition forces specially trained in mountainous environments are better trained overall.[35]

Acclimatization training. Mountain troops should be stationed at high altitudes to maintain a high standard of physical fitness and acclimatization. Before World War I, German alpine troops were stationed in the Bavarian Alps.[36] In Transcaucasia at 6,500 feet, Soviet troops occupied ideal terrain to train soldiers.[37] Pakistan and India maintain a large number of troops on the line of control in Kashmir. This automatically provides ample opportunities for ac-climatization. The US 10th Mountain Division, however, is stationed at Fort Drum, New York, which is not in high mountains.

Although troops need to be acclimatized for any kind of mountainous terrain, the duration depends on the altitude at which the unit must operate. Acclimatization for mountains below 13,000 feet takes three to four weeks. Pakistan troops train by bivouacking at high altitudes and conducting routine administrative activities and route marches. Each week they conduct hill climbing at increasingly higher altitudes to increase their ability to function. The rigorous training also helps identify soldiers who have medical problems.

Acclimatization for higher altitudes is rigid, and the length of training cannot be shortened without serious consequences. Although the pattern of training remains the same, troops are trained at an altitude of between 8,000 and 10,000 feet for two weeks, followed by one month's training at 11,000 feet. The troops conduct route marches, fire weapons, climb rocks, and cross crevasses. They then move in stages from 13,000 feet to forward posts at heights up to 21,000 feet. The basic principle is to bivouac one night for about every 3,000-foot increase in altitude. Troops returning from leave must repeat the process.

Because the US 10th Mountain Division is not stationed in a high-altitude environment and its soldiers are not acclimatized, deployment to high mountains would require an additional two to three weeks of acclimatization. This also applies to units that have conducted mountain-warfare training but are not currently stationed at high altitudes. Physical conditioning is essential to mountain combat in Afghanistan.[38]

Leadership training. Leadership training is extremely important for mountain warfare. During World War II, German alpine troops displayed strong leadership traits based on their culture of *auftragstaktik*.[39] In the Pakistan Army, mountain training is considered part of overall training. The Soviet Army recognized the critical importance of junior-leader initiative during its experiences in Afghanistan.[40] The US Army focuses on leadership, but it does not have a package designed specifically for training at different levels of leadership in mountainous environments. Mountain leadership training should be based on the unique characteristics and demands placed on leaders.

Junior leader training requires initiative, personnel management, and mental toughness. In the Pakistan Army, most of these skills are learned through experience and exposure to tough environments. During mountain-combat deployment, junior-leaders often conduct patrols, lead expeditions, and direct command posts, despite sometimes having to endure heavy artillery shelling and adverse living and weather conditions. The US Army Mountain Warfare School emphasizes gaining mountaineering skills rather than training combat leaders. The US Marine Corps Mountain

Warfare Center has a mountain leader course designed for junior leaders, but it is heavily skills-oriented.[41]

Mountain leader training should begin with an introduction to issues and problems unique to mountainous terrain. Practical exercises and historical case studies increase leader awareness. Training outdoors in command positions with specific tasks, such as navigation, patrolling, raids, and ambushes, should be conducted at altitudes above 8,000 feet. Simulations that force junior leaders to make tough choices between their soldiers' physical capabilities and mission accomplishment are essential to mountain leader training.

Many senior leaders consider themselves well-equipped to plan and conduct operations in any environment, but commanders who have served in the mountains as young leaders are far better at understanding and planning for such environments. Wargames and live exercises are valuable tools to help senior leaders understand mountain warfare. Live exercises illustrate human limitations in such an environment.

Individual/team training

Most armies from countries with mountainous terrain have well-established training institutions. Location of training institutions is an important consideration for mountain training. The Pakistan Army's High-Altitude School, at Rattu in Northern Kashmir, is an ideal location on the confluence of the Hindukush, Himalayas, and Karakorum ranges. The school conducts training throughout the year and includes mountain climbing on peaks ranging from 15,000 to 20,000 feet and survival on glaciated terrain and in snowy and icy conditions. The Indian Army's high-altitude warfare school is at Gulmarg, which is at 8,000 feet.[42] The US Army Mountain Warfare School is located in Vermont [which is at 560 feet], while the USMC Mountain Warfare School is located in Bridgeport, California, which is at 9,000 feet.

The purpose of these training institutions is to train individuals to survive and take advantage of the extreme terrain and weather conditions in the mountains. Physical conditioning is the first prerequisite of mountain warfare. Training should be progressive, starting with light physical exertion followed by route marches and mountain climbing, culminating in test exercises in difficult terrain. Most schools have similar programs for this purpose. Although training individuals at heights up to 10,000 feet can achieve a great deal, to develop high-altitude skills, some training should occur above 13,000 feet.

The ability to navigate and move across difficult terrain builds confidence and enables soldiers to plan and execute maneuvers across seemingly impenetrable and inhospitable terrain. The small-unit mountain operation exercise the US Army Mountain Warfare School conducts is a good example.[43] Participants must navigate to six different sites within a set time. The exercise incorporates various skills soldiers need to complete actions successfully in a time-compressed, competitive environment. Some skills frequently required during mountain warfare include using rope bridges and vertical haul lines and medical evacuation. Mobility in winter has several prerequisites, including using snowshoes, skiing, climbing ice, crossing crevasses, and detecting avalanche hazards.

A key training objective in mountain training is properly using winter clothing, weapons, and equipment and recognizing and preventing cold-weather injuries. Such skills are especially important for officers and noncommissioned officers who must enforce these practices. Lectures, demonstrations, and practical experience can help address these problems. The Pakistan Army's standard training procedures cover most safety issues, such as frostbite prevention, high-altitude sickness, and pulmonary and cerebral edema.

The Army's Mountain Warfare School teaches winter sustainment using the Akhio tent and stove group. The Akhio sledge contains a 10-man arctic tent, a diesel-fired stove, fuel, and other basic supplies. Kashmir tents have a short lifespan because of the wear and tear from blizzards and heavy snowfall. Stone structures and synthetic igloos are the preferred structures. In noncontiguous and nonlinear battlefield environments, soldiers must build protective shields around winter shelters to avoid becoming targets for raiding parties and artillery fire.

Once trainees understand mountainous terrain and its effects on combat, the next step is to conduct small exercises involving patrolling, raids, and ambushes. These exercises should incorporate mountaineering skills in situations that tests trainees' abilities to modify traditional tactics to mountainous terrain. These exercises build leadership skills, initiative, flexibility, and team spirit. Although no opposing force (OPFOR) exists in the Pakistan Army's High-Altitude School or at the US Army Mountain Warfare School, an OPFOR is necessary for creating a realistic environment and developing mountain warfare skills.

US Army Mountain Warfare School instructors are topnotch mountaineers. However, most do not have actual mountain-combat experience.[44] Posting officers who served in Afghanistan to the Mountain Warfare School might address this problem.

Collective training

Collective training is an opportunity to test units and formations in actual mountainous environments, reinforcing and building on skills gained through acclimatization and individual training and allowing commanders to check the viability of their assumptions and plans in a realistic setting. Synchronization and coordination between fighting and supporting arms and among all the battlefield operating systems are also key elements of this training.

Collective training in winter and summer environments is a regular part of the Pakistan Army's mountain training. Because altitude is an important consideration, reserve units train at heights equivalent to those at which they are expected to fight. Training is primarily mission-centric, based on the nature of tasks assigned to the units, and includes offensive and defensive tasks and small-unit actions.

The US Army does not conduct collective training in mountain warfare; it focuses more on survival training rather than high-altitude combat.[45] The USMC conducts infantry battalion training, but the training does not include artillery, engineers, aviation, or other supporting arms.[46] Considering the unique requirements of mountainous and high-altitude en-vironments, these can be serious limiting factors for coordinating and synchronizing the combined arms fight and can easily lead to faulty planning and wrong assumptions about each other's capabilities and limitations.

Operation Anaconda demonstrated that fighting in the mountains is not a special operation or exclusively an infantry domain.47 Mountain warfare involves logistics, aviation, artillery, communications, and air assets. With the level of sophistication in these branches and services, there is an even greater need for collective training in order to use their unique characteristics fully.

Branch-specific training

All branches and services need to train for mountain combat to understand the capabilities and limitations of their equipment. Aviation is critical to mobility, timely logistics, and precision firepower. Pilots should be well trained in mountain flying and in understanding an infantryman's problems in mountainous terrain. The Pakistan Army's 8th Aviation Squadron supports operations in Kashmir. Pilots have hundreds of hours of combat flying experience and understand the mountainous environment.

With the enhanced capabilities of Apache helicopters to acquire and engage targets beyond visual range, US Army pilots, in conjunction with ground troops, need to practice firing in the mountains. The Russians recognized the need for close coordination between aviation and ground troops during their war in Afghanistan.[48] The US experience in Afghanistan highlighted the need for attack aviation to train with Special Operations Forces and to practice using night-vision devices.[49] Pilots for cargo and troop-carrying helicopters also need to train in mountains in various weather conditions. High-altitude training combining attack and cargo helicopters is essential for high-altitude combat.[50]

The US Army has a variety of sophisticated communication equipment. Although some equipment works well in the mountains, some requires improvisation and alternatives because FM communications are often ineffective at high altitudes and distances impeded by mountainous terrain.[51] Shifting retransmission stations and using equipment in various weather conditions is essential to providing a variety of options to communication providers and users.

Engineers, who are key to mobility, countermobility, and survivability, must also train and work in high-altitude conditions. Constructing shelters, laying minefields, providing clean water supplies, and constructing bridges and roads require different considerations in the mountains. Training in a mountainous environment is the only way to ascertain the type and quantity of materials and equipment needed.

Artillery units need to train in mountains to ensure optimum fire support under all circumstances and all weather conditions. Selecting gun positions and shifting and readjusting guns by air, as well as ground transport, to support various tactical contingencies require training and experience, as do observation and fire direction.

In mountainous environments, CSS elements also need to learn how best to use trucks, aircraft, porters, and mules. Training under real mountain conditions helps them identify the differences in logistical calculations for mountainous environments. Other issues, such as protecting logistic bases, are equally important. The Mujahideen often successfully attacked and destroyed Soviet logistic bases in Afghanistan.[52]

Doctors and medical staff also need special training in recognizing and treating high-altitude-related injuries and illnesses. In Kashmir, doctors are fairly confident in dealing with them. During World War II, medics accompanying German mountain troops were experts at treating frostbite, snow blindness, and other problems.[53] The Soviet Army instituted more than 100 hours of training through a special course for doctors and their staffs.[54]

Contemporary High-Altitude Operations

The US Army had a distinguished history of mountain operations during World War II. The 10th Mountain Division proved its utility as an elite mountain-trained force during the Italian Campaign. More than 50 years later, Operation Anaconda in Afghanistan highlighted the continued need for specialized training in mountain warfare. The Army has embarked on an impressive Transformation intended to meet the challenges of the 21st century; expert mountain troops available for deployment on a short notice should be a part of this Transformation.

Major Muhammad Asim Malik, Pakistan Army, received an M.M.A.S. from the US Army Command and General Staff College. He also attended the Command and Staff College, Quetta, Pakistan. He has served as Brigade Major of an Infantry Brigade; an instructor at the Pakistan Military Academy, Kakul; a company commander of an Infantry Company, Siachen Glacier, Kashmir; aide-de-camp to the corps commander; and a platoon commander and adjutant of the 12th Baloch Regiment, Infantry, Kashmir.

Notes

1 James Lucas, *Alpine Elite: German Mountain Troops of World War II* (London: Janes, 1980), 13, 130.
2 US Army Field Manual (FM) 3-97.6, *Mountain Operations* (Washington, DC: US Government Printing Office [GPO], 28 November 2000).
3 CPT John Clearwater, *Above and Beyond*, on-line at <www.pakmil/com/army/insti/highalti.html>, 1, accessed 12 September 2002.
4 Bruce C. Patton, *Cold Casualties and Conquests: The Effects of Cold on Warfare*, 23, on-line at <www.Armymedicine.Mil/history/Borden/medaspofharshenvrnmnts>, accessed 29 November 2002.
5 FM 3-97.6. At 18,000 to 22,000 feet, the glaciated area in northern Kashmir is considered the highest battleground in the world.

6 Brigadier Ghazanfar Ali and A. Ghani, *Siachen: The World's Highest Battlefield*, on-line at <www.Pakdef.info/pakmil/army/siachen/>, 3, accessed 2 December 2002.
7 CPT John R. Ballard, "Training an Arctic Raid Force," *Marine Corps Gazette* (February 1987): 64; LTC Salman Beg, "Operations in Glaciated Areas," Pakistan Army Journal (Spring 1994): 3.
8 Kevin Fedarko, "War at 21,000 Feet," *Outside Magazine* (April 2002): 41, 44.
9 LTC John E. Sray, *Mountain Warfare: The Russian Perspective* (Washington, DC: GPO, March1994), 20, on-line at <www.fmso.leavenworth.army.mil/fsmopubs/issues/ mountain>, accessed 12 September 2002.
10 Ali and Ghani, 4.
11 Ibid.
12 Clearwater, 2.
13 Lester W. Grau and William K. Jorgensen, "Medical Implications of High Altitude Combat," *US Army Medical Journal* (April 2002): 1; Ali and Ghani, 3.
14 Fedarko, 50.
15 LT Craig M. Banull, "High Altitude Medicine: Case Report," *Navy Medicine* (January-February 2000): 27.
16 Ibid., 48.
17 Sray, 1.
18 Lucas, 196.
19 FM 3-97.6, 2-15.
20 Ballard, 63.
21 MG D.K. Palit, *War in High Himalaya* (London: Lancer International, 1991), 205.
22 Sray, 8.
23 George J. Mordica, *High Altitude Operations*, on-line at <http//call.army.mil/ products/ trngqtr/tq4-02/mordica>, accessed 2 December 2002, 7.
24 Robert F. Baumann, "Russian Soviet Unconventional Wars in the Caucasus, Central Asia, and Afghanistan," *Leavenworth Papers 20*, Fort Leavenworth, Kansas, November 2000, 141.
25 Ahmed Ashfaq, *Fangs of Ice: The Story of Siachen* (Rawalpindi: Pak American Commercial, 1991), 25.
26 FM 3-97.6, 3-29.
27 Sray, 16-18, Grau and Ali a Jalali, *The Other Side of the Mountain: Mujahideen Tactics in the Soviet-Afghan War* (Washington, DC, GPO), 187.
28 Gerhard Sheppe, *Mountain Warfare in Europe* (Kingston, Canada, 1983), 37.
29 Fedarko, 50.
30 Lucas, 130.
31 Ali and Ghani, 3; FM 3-97.6, 5-7.
32 Mordica, 9.
33 Ibid., 5-13.
34 Ibid., 4-5.
35 Ibid., 4.
36 Ibid.
37 Sheppe, 35.
38 Mordica, 4.
39 Lucas, 197.
40 Sheppe, 35.
41 Grau, interview by author, January 2003.
42 *Indian Armed Forces, Training Institutions*, on-line at <www.indianarmedforces. com/def/ army/def9.html>, 6, accessed 12 September 2002.
43 Grau and Jorgensen, 5.
44 Grau interview.

45 Major John G. Bechtol, "Fighting the Cold: The Need for Standing Cold Weather Combat Capabilities," unpublished article, *Naval War College*, 4 February 2002.
46 Grau interview.
47 Mordica, 4-5.
48 Sray, 20.
49 Mordica, 4-5.
50 Ibid, 8.
51 Ibid, 9.
52 Grau, *The Bear Went over the Mountain: Soviet Combat Tactics in Afghanistan* (Collingsdale, PA: Diane Publishing Co, March 1996), 128.
53 Lucas, 210.
54 Grau and Jorgensen, 5.

SMALL UNIT

3

Forbidden Cross-Border Vendetta: Spetsnaz Strike into Pakistan during the Soviet-Afghan War

LTC Lester W. Grau, US Army (ret.) and Ali Ahmad Jalali

This article first appeared in the December 2005 *Journal of Slavic Military Studies*

Lieutenant-Colonel Babushkin wanted revenge. The Mujahideen guerrillas maintained a large base camp in his 15th Spetsnaz Brigade area of operations and the Soviet command in Kabul would not let him go get it. The Mujahideen used their border base at Krer to launch forays against him and resupply Mujahideen forces throughout northeast Afghanistan. In February 1985, the Mujahideen "Black Storks" unit had massacred one of his Spetsnaz companies. Only Captain Oleg Mart'yanov, the deputy commander, and a wounded RTO survived. Then in August 1985, Captain Bykov and 50 of his men were caught during a reconnaissance of Krer's well-prepared forward defenses and took more casualties.

The reaction of 40th Army was gutless. They forbad any combat within five kilometers of the border. They were worried that Soviet forces might stray across the Pakistan border and create an international incident. The Mujahideen crossed that border daily and no one seemed to give a damn. Now, however, the Soviets finally had a "tongue"—a captured Mujahideen who served in Krer and who, with some persuasion, provided good intelligence on the personnel strength, weaponry and laydown of the defending Mujahideen regiment. The time to settle Krer was now, and those bastards in Kabul were playing international niceties. The frigid January wind howling down the Kunar valley did nothing to cool his desire for revenge. (See map on page 2 of this PDF.)[1]

Krer–Mujahideen Bastion on the Kunar River

By January 1986, the Soviet-Afghan War had settled into a struggle to strangle each other's logistics. The Mujahideen targeted the Soviet's long convoy route and trucks hauling supplies from the Soviet Union to the major garrisons in Afghanistan. Since the Mujahideen depended on the rural population for food and shelter, the Soviets attacked the rural population's villages, standing crops, orchards, granaries, animal herds, windmills and irrigation systems with bombers and artillery to drive them off their land. This ruthless approach forced over seven million (of 17 million) people in Afghanistan to become refugees fleeing to the cities of Afghanistan or the refugee camps in Pakistan and Iran. It also forced the Mujahideen to carry their own rations with them on operations. In order to stay in the fight, the Mujahideen were forced to create a logistics network of bases, dumps and caches in order to sustain their forces. Krer was a major logistics base supplying Mujahideen forces in Kunar Province and northeast Afghanistan.

The Krer base complex was located on the route from Bajuar, Pakistan, a Mujahideen supply distribution point and camp. Krer was garrisoned by the Asama Ben Zaid Regiment. It was usually under-strength but could grow to some 400 for an upcoming battle.[2] (Soviet intelligence estimated the regiment's present-for-duty strength at 500 and misidentified it as the Abdul Vakilya Islamic Regiment). The regiment belonged to the Sayyaf faction and was commanded by Commander Assadullah.[3] It was well armed with mortars, recoilless rifles, and DShK and ZGU heavy machine guns. Krer base was located some 20 kilometers southwest of the Afghan town of Asadabad at the head of Krer canyon on the heights of Spina ridge and Spina Tsoka peak. The approach to Krer from Pakistan is easy and gradual, but the approach from the Afghanistan side is difficult. First, the broad Kunar River has to be crossed. A bare plateau, some 600 meters [1969 feet] above sea level dominates the river crossing. Then mountains rise sharply from the plateau until they reach some 2000 meters [6562 feet] above sea level along the Spina ridge. The first 1000 meters of the mountains are bare and then larch, aspen and juniper forests and alpine meadows begin and continue to the top.

The 15th Spetsnaz Brigade was headquartered in Jalalabad and commanded by LTC Babushkin.[4] The brigade's 334th Spetsnaz Battalion was garrisoned in Asadabad and was commanded by Captain Grigori "Grisha" Bykov.[5] The brigade's 154th Spetsnaz Battalion was garrisoned in Jalalabad and was commanded by Captain Roman "Roma" Abzalimov.[6] The brigade drew aviation and artillery support from the 66th Separate Motorized Rifle Brigade that was also stationed in Jalalabad. Often, the brigade was augmented by forces from the Democratic Republic of Afghanistan (DRA) army, Sarandoy (armed police) and KHAD (KGB). The brigade's primary mission was to interdict Mujahideen supplies coming in from Pakistan.

The Reconnaissance Raid

LTC Babushkin needed more information before launching a major effort against Krer, so he planned a reconnaissance raid against Krer's forward defenses. He drafted a plan to establish ambushes at the fording sites along the Kunar river close by the Krer canyon approaches. He sent this plan up through channels and the 40th

Army staff in Kabul approved it. LTC Babushkin then dispatched a raiding force from the 334th Spetsnaz battalion to the ambush sites. The raiding force consisted of two company-command groups with about 45 men each. The two companies trudged through the winter night to the ambush sites. Once on site, they radioed their battalion Combat Control Center and reported that they had been in contact and were pursuing the fleeing enemy into Krer canyon. In fact, there had been no contact. The raiders moved onto the plateau. They climbed the heights until they discovered two forward Mujahideen posts. The company commanders talked it over and decided to raid the posts. They reorganized their force into three groups—two raiding groups and a support group that would stay on the mountain ridge to provide covering fire and support the raiding groups' withdrawals.

As the Spetsnaz neared the posts, it became obvious that one of the posts was much harder to reach due to the rugged terrain. A simultaneous raid on both sites was impossible. The company commanders conferred on the move and altered the plan. One group would first attack its post over the more-accessible approach route. The second raiding group would wait for the enemy security element in the second post to get out of their bunkers to determine what was going on at their neighboring post. The first group attacked and overran its post. As expected, the Mujahideen in the neighboring post occupied their positions, but when they could not detect a threat, they climbed out of their positions to try and see what was going on to their flank. At that moment, the second group opened up on the Mujahideen and then overran their post.

The Mujahideen posts were well laid out—by the book. There were deep bunkers and storage areas for arms, ammunition and food. They were prepared for sustained, unsupported combat and were equipped with radios and telephones. The raiders held the posts for a total of ten minutes. They grabbed enemy weapons and blew up those enemy weapons they couldn't carry. The support group fired on the remaining Mujahideen forward security posts in order to draw their fire while the raiding groups withdrew into the darkness. The raid was successful, developed good intelligence information and proved that there was a way to get around those sanctimonious meatheads in Kabul who were so nervous about border fighting. It was time to begin planning for the main event.

Getting ready for the Main Event

The information from the raid and from the Mujahideen "tongue" helped develop a picture of how Krer was defended. LTC Babushkin began planning a feint and end-run designed to destroy Krer base. The attack would go in on the afternoon and night of 30 March 1986. The 334th and 154th Spetsnaz battalions would conduct the attack. They would be supported by a platoon of 122mm D-30 howitzers and a BM-21 multiple rocket launcher from the 66th Separate Motorized Rifle Brigade. DRA and KHAD troops would accompany the raid.

The staffs of the two battalions began working together on details of the plan. Grisha Bykov's 334th Spetsnaz Battalion would travel from Asadabad to the District

Capital of Sarkani on armored personnel carriers. There, they would cross the river, move to the mountain and then dismount and climb. Their personnel carriers would cover their advance as they climbed up the same northern crest where they had earlier conducted the reconnaissance raid. They would advance up this crest while drawing Mujahideen attention to their approach. Roma Abzalimov's 154th Spetsnaz Battalion would move from Jalalabad to the ferry site on armored personnel carriers. The artillery would accompany them. The carriers and artillery would remain on the western bank. The raiders would ferry across the river and secretly advance along the southern ridge while the Mujahideen were focused on the northern advance. In the morning, both battalions would link-up on Spina ridge near Spina Tsoka peak. They would destroy Mujahideen weapons, ammunition and fortifications as they advanced. Once they seized the ridge and linked up, they would remain for the day, destroying the base, and then withdraw under the cover of darkness.

The battalions conducted mission training, issued ammunition and inspected equipment for the raid. Unfortunately, there was an outbreak of hepatitis in the 154th Spetsnaz battalion.[7] Even after drafting cooks, clerks and other battalion support personnel, the battalion could only muster 150 combat-effective personnel for the mission. Lieutenant Oleg Mart'yanov's 3rd Company was the strongest with some 70 personnel. The 1st Company was second strongest and it was lead by Lieutenant Udovichenko, the deputy company commander. The 2nd Company was down to two platoons and moved with the battalion headquarters.

The Raid Begins in the North

The weather was still cold and the mountaintops were covered with snow. Commander Assadullah, the Krer commandant, could see a column of Soviet and DRA vehicles approach Sarkani in the late afternoon. They moved through the town and across the Nawabad Bridge. They headed toward his northern ridge—the main approach route into Krer base and the expected route of enemy attack. Heavy fighting broke out as the Soviets dismounted and began to attack the forward security positions under the covering fire of their armored personnel carriers. Darkness fell and fighting continued. The 334th Battalion had a local guide, but the battalion got lost in the darkness and finally ended up climbing a flank of the ridge. Getting lost was lucky since the Mujahideen concentrated murderous heavy machine gun and recoilless rifle fire against the expected Spetsnaz route of advance. The Spetsnaz approached the Mujahideen positions from the flank and systematically took them out with hand grenades. Still, both sides took heavy casualties as they slugged it out in the night.

One of the Mujahideen groups that the 334th bumped into was not from the regiment guarding Krer. Ahamadullah Wejdani and his group of 16 Mujahideen were returning from Pakistan to their base in the Pech Valley of Kunar Province. They were armed with two PK machine guns and 15 AK-47 rifles. They were carrying a large load of ammunition and a battery-operated megaphone. They had tried to cross the river earlier, but the ferry operator warned them of DRA and Soviet military activity in the area. They decided to return to Pakistan and wait for things to calm down. On the way back to Pakistan, they formed a defensive perimeter and camped for the night. About ten PM, they heard noises that sounded like a stampede of wild hogs. They jumped to their feet and, in a flash of lightning, saw a large group of soldiers below them climbing in their direction.

Ahamadullah Wejdani put his men in position to open fire on the soldiers during the next lightning flash. Moments later, the target was illuminated by lightning and they opened up on the Spetsnaz. The Spetsnaz returned fire and the Mujahideen ceased fire and withdrew up the hill. They lost their megaphone during the withdrawal. After the Mujahideen climbed some 300 meters, they settled in to repeat the process. They heard the squeal from their megaphone and realized that the Spetsnaz were in their old positions and playing with the megaphone. They opened fire on their old position and started to climb again. They systematically withdrew up the mountain, firing at the Spetsnaz and then moving out of the path of return fire.

The Raid Begins in the South

The 154th Spetsnaz Battalion left Jalalabad on the night of 29 March and arrived at the ferry-crossing site in the morning. The battalion dismounted and began to cross. The armored vehicles and artillery stayed on the western bank. The ferry was old and it took four hours to get the raiding party across. Then the group started to climb the plateau and moved toward the mountains. The second platoon leader of the 1st Company had to be replaced since he had come down with hepatitis. By nightfall, the southern raiding group started climbing the mountain. It was tough going and the group that was having the most trouble was the 20-man rear guard composed of DRA and KHAD troops. They were out of shape and refused to keep up. When the rear guard commander, Lieutenant N. Zubkov, radioed his problem to the battalion commander, Captain Roma Abzalimov replied "Abandon them. Screw them." When Zubkov relayed this to the reluctant warriors from the KHAD, they realized that being abandoned in Mujahideen territory might be unpleasant and suddenly were able to keep up with the battalion.

The terrain was rugged and the climb was tough. At several places, the only way the soldiers could climb up was by standing on each other's shoulders. Thanks to the attack by the 334th Battalion, the 154th Battalion was able to climb to the top of the mountain undetected. They reached the Spina crest shortly before dawn. The company commanders and battalion commander were deciding on how to seize and occupy the area when a loudspeaker broke the early calm. The loudspeaker was waking the faithful and calling them to morning prayers. The 3rd Company immediately assaulted a Mujahideen firing position they discovered during the ascent.

The battalion moved across the Spina crest, seizing well-constructed, amply-supplied Mujahideen fighting positions and digging in. By 0400 hours, the 154th Battalion controlled the Spina crest and could fire down at the Mujahideen trapped below them. They overran the bases and supply depot and captured a lot of ammunition and many Mujahideen heavy machine guns and RPGs. The 1st Company moved onto the heights where it could dominate all the approaches into the camp from Pakistan. The company began to dig in some 700 meters inside Pakistan. The Spetsnaz controlled the area. Ahamadullah Wejdani's group was still fighting a systematic retreat but was now out of ammunition and could see the signal rockets of the Spetsnaz above them. He gathered his men and withdrew through a side canyon and headed into Pakistan.

Counterattack

Dawn broke and the winter sun warmed the troops of the 154th Battalion. Another lieutenant fell to hepatitis and had to be replaced. Around 0730 hours, 40 Mujahideen came strolling toward the 1st Company positions inside Pakistan. They were chattering and relaxed and many of them were at sling arms. When they were about 70 meters from the 1st Company, the Spetsnaz opened up on them. Some 15 fell and the rest jumped behind the shelter of some rocks. About 40 minutes later, some trucks began to move toward them from the town of Bajuar. When the 154th Battalion overran the Krer bases and supply depot earlier, Commander Assadullah had hurriedly crossed over to the Mujahideen camps in Bajuar to raise a relief force. That relief force was now loaded on the trucks. The Spetsnaz called artillery fire on the Mujahideen as they began to dismount from the trucks into an assembly area. The artillery fire slowed the Mujahideen down, but the Mujahideen counterattack was beginning. Soon the Mujahideen assault fire was so thick that the Spetsnaz had to hold their assault rifles up over the rocks and fire back blindly. The Mujahideen approached closer to the Spetsnaz. Lieutenant Osobenko continued to adjust artillery but it was not falling close enough to allow the Spetsnaz to break contact. Osobenko finally called the artillery on top of his own position.

After a brief radio argument with the brigade commander, the rounds slammed onto the Spetsnaz position. The surviving Spetsnaz withdrew pell-mell. The Mujahideen now controlled the high ground and were firing directly into the command post of the 154th Battalion. It was now 1045 hours and the 1st and 2nd Companies were combat ineffective. Radio calls for help from the 3rd Company went unanswered. The 334th Battalion was still below the 154th Battalion and there were an increasing number of Mujahideen moving into the gap between the two battalions. The Brigade Commander finally requested helicopter gun ship and medevac assistance—disclosing his unauthorized raid to the high command. But as the first medevac helicopter approached the 154th Battalion, it was unable to land due to the intensity of the Mujahideen fire.

Enter the Gun ships

Commander Assadullah's counterattack was going well. While some of his force overran the defending 1st Company of the 154th Battalion, the bulk of his force infiltrated through concealed approaches into Krer. His gunners occupied the high ground and fired down into the disjointed Spetsnaz. Commander Assadullah's six-man security group bumped into the command group of the 1st Company and killed a Soviet officer during the fierce AK and grenade fight. The 154th Battalion was falling apart and only air strikes and artillery could save the pieces.

Finally, Soviet helicopter gunships arrived overhead. Lt. Osobenko, who had survived the artillery strike on his own position, was now inside an adobe building with other Soviet wounded. Mujahideen were on the roof tossing F-1 hand grenades at the building doorway. LTC Babushkin directed Lt. Osobenko to adjust helicopter gunship fire. Lt. Osobenko called the helicopter flight leader and asked him them to hit various targets. Instead, the pilot answered "I am not allowed to because of the international border." Then the helicopters made a dry run over the area, but did not fire. They did this six or seven more times and then, after a half hour of dry runs, a quiet, ice-cold voice spoke over the radio "I am tail number 25 and I am ready to fire. Give me the target."

Lt. Osobenko directed the gunship runs. They first cleared the Mujahideen from the roof of the building in which Osobenko and the wounded were sheltering. Then the helicopters flew repeated low-level gun runs at the Mujahideen. The helicopters stayed on station until darkness. The surviving Spetsnaz began dragging their dead and wounded downhill as they moved to link up with the 334th Battalion.

Policing the battlefield

The helicopter gunship strikes and artillery fire had saved the 154th Battalion from certain annihilation, but their trial was not over. Two companies of the 334th Battalion pushed up the mountain to link up with the 154th. These companies helped carry out the dead and wounded of the 154th. The nearest spot that helicopters could set down that was not controlled by the Mujahideen was on hill 1917—about ten kilometers away. This was not a regular landing zone, rather a rock ledge where a helicopter could touch down with one wheel on the ledge while ground troops threw the dead and wounded into the troop compartment. As the night passed, one of the companies of the 334th went back up the mountain to look for the dead and wounded.

The morning of the 1st of April dawned as the company pushed forward. Soviet air and artillery strikes were pounding the Krer region. Later in the morning, Soviet helicopters landed the air assault battalion of the 66th Separate Motorized Rifle Brigade from Jalalabad. The troops of the 66th and the 154th were able to find and evacuate more Soviet dead, wounded and missing. Two patrols of the 334th strayed into Pakistan and discovered a Pakistani border security post apparently abandoned by the Pakistani border guards with the advent of the fighting. Finally, on the morning of 2 April, the Soviets abandoned the search for their two remaining missing Soviet Spetsnaz and withdrew completely from the fight.

Aftermath

Mujahideen casualties among the Krer regiment were 33 killed and 40 wounded (about 25 percent of the regiment's pre-battle strength). Exact casualty figures from the Mujahideen counter-attack force are not known. Soviet casualties have not been released. Two Soviets hid in one of the supply caves after the general Soviet withdrawal and were eventually killed following a prolonged fight. The Soviets never recovered their bodies. The Mujahideen captured some 60 Soviet small arms and felt that Soviet casualties were at least 60. Indications are that they were much higher.

The Soviet 40th Army forbad any future attacks on Krer (although they attacked again in December 1987). The over-blown Soviet estimates of Mujahideen losses were over 300 casualties and they also reported the lurid, but fictitious, account of the Mujahideen execution of Commander Assadullah and his deputy for letting the base be overrun. Soviet after-action reports also incorrectly stated that Gulbuddin Hekmatyar, the leader of the major Islamic Party faction, personally led the fighting on 31 March.[8]

LTC Babushkin and his officers were quickly called on the carpet for an official investigation of the cross-border fight. The colonel and his officers were waiting for a helicopter to Kabul in a Jalalabad airfield hanger. Lieutenant Osobenko approached some aviators and said "Guys, who's the fag who flies tail number 25?"

The aviators stared back at him and LTC Tseloval'nik answered "That's my bird. Do you have a problem with it?"

"Yeah, what's the idea of leaving us stranded for a half-hour while you made all those dry-fire runs?'

LTC Tseloval'nik replied, "As soon as we returned to this airbase, we began reloading and refueling our aircraft. Uniformed KGB officers immediately approached me and said 'Explain to us, Lieutenant Colonel, who gave you the right to conduct combat beyond the border?' I looked back at them with astonishment and denied doing so. The military prosecutor who was with the KGB stated 'There was a group across the border and you provided fire support for them.' 'That cannot be,' replied. 'Here, listen to the flight recorder tapes.' They listened to our tapes and heard your requests and our denials over and over again." The aviators had played this charade for a half hour, then shut off the tapes and provided fire support.

The inquiry found against LTC Babushkin and he was relieved of command and sent back to the Soviet Union. Commander Assadullah rebuilt the Krer base and it was soon supplying the Mujahideen in northeast Afghanistan again. The Pakistani *Strategic Studies Review*, which provided detailed coverage of the war reported in April on the Krer fighting. "Soviet forces launched air-cum-ground attack on Mujahideen base in Krer area killed 26 Mujahideen destroyed their entire armament and lost 42 men after 15 hours fighting March 26. Mujahideen killed 70 Kabul and 50 Soviet troops and lost 42 men in their bid to break Soviet-Kabul encirclement of Soran base in Krer area of Sarkani District March 28-31.[9] Mujahideen repulsed Soviet attack after hours of occupation of their base in Krer after inflicting heavy losses and capturing three Soviet troops during 48 hours fighting March 30-31." After all the Soviet concern about cross-border operations, the Pakistani authorities were either unaware that the Soviets had crossed their international border or did not want to make an issue of it.

Krer was a major stationary depot and had to be defended—the antithesis of guerrilla war. To hold the base, the Mujahideen built their defenses around heavy crew-served weapons and well-positioned permanent fighting positions. These protected them from Soviet aircraft and artillery fire. But Krer was not protected from well-trained Soviet light infantry who moved at night. The Mujahideen felt that Soviets did not fight well at night and would not fight at long distances from their armored personnel carriers. The Spetsnaz were the exception and they succeeded in overrunning Krer. However, since they did this in secret, they were unable to mass Soviet fire power in support until it was too late. The Spetsnaz vendetta ended badly—as uncontrolled actions by elite forces often do.

Notes

1. Source of the Soviet information in Sergei Kozlov, "Karera: Novyy vzglyad" [Krer: A New Look], Soldat udachi [Soldier of Fortune], Number 7, 1997, pages 4-9 and 41. His account is based on interviews with four officer participants. The map is based on a Russian map from this article.
2. Asama Ben Zaid was a close companion of the Prophet Mohammad and one of the prophet's military leaders. At one point, Asama Ben Zaid fought the Byzantines.
3. Commander Assadullah gave an interview to the authors. His account of the battle is in Ali Ahmad Jalali and Lester W. Grau, The Other Side of the Mountain: Mujahideen Tactics in the Soviet-Afghan War, Quantico: USMC Study DM-980701, 1998, 327-330. The source for other Mujahideen information in this article is an unpublished interview conducted by Ali Jalali in the summer of 1998 with Ahamadullah Wejdani, a Mujahideen Commander from the Kahlis faction.
4. The 15th Spetsnaz Brigade codename was the 150th Spetsnaz Brigade.
5. The 334th Spetsnaz Battalion was also known as the 334th Separate Spetsnaz Detachment and the Asadabad battalion. Its code name was the 500th Spetsnaz Detachment or battalion.
6. The 154th Spetsnaz Battalion was also known as the 154th Separate Spetsnaz Detachment. Its code name was the 100th Spetsnaz Detachment or battalion.
7. Viral hepatitis was common among Soviet forces and often incapacitated entire units. It was a product of poor field sanitation–particularly the failure of cooks to wash their hands thoroughly after defecating. During 1980, the entire 5th Motorized Rifle Division in the Western corridor of Afghanistan was rendered combat ineffective by hepatitis. See Lester W. Grau and Dr. William A. Jorgensen, "Medical Support in a Counter-Guerrilla War: Epidemiologic Lessons Learned in the Soviet-Afghan War," U.S. Army Medical Department Journal, May-June 1995.
8. Krer was garrisoned by fighters from the Islamic Union for the Liberation of Afghanistan faction.
9. Strategic Studies Review, April 1986, Islamabad, Pakistan.

4

Small Secrets of Great Mountains

LTC Lester W. Grau, US Army (ret.) based on the work of Ikram Nazarov

This article first appeared in the May-June 2010 *Infantry Magazine*

Usual military mountain training focuses on individual climbing skills. How to place pitons, how to rappel down slopes, how to tie knots, how to ascend and descend steep rock faces or cross loose fields of scree (rock fragments) or shelves of snow. Armies, however, do not fight as individuals. They fight as units and so mountain training should concentrate on unit movement and unit fighting in the mountains. Units should learn to climb, to fire and to maneuver by teams, squads and platoons. The people who support the infantry— the gunners, the sappers, the logisticians and the transporters also need training in mountain movement and mountain com- bat skills. It is particularly beneficial when they accompany the Infantry for this training.

Ikram Nazarov is a mountaineer instructor/trainer with more than forty years experience in military mountaineering. He was involved in training Soviet units for mountain combat during the Soviet-Afghan War and has trained Russian units for the mountains of Chechnya. Now he is working with the Army of Uzbekistan imparting mountaineering skills. He published the following article in the May 2008 issue of the Russian Armeyskiy sbornik *(*Army Digest*).*[1] *The article demanded translation so that it can reach a wider audience. Although the Russian Army and the US Army train and fight differently, there is much in it that is of use to the American mountain soldier.*

In the mountains, the time set to depart on a mission and to return to camp is law. If a group of mountain soldiers has not re- turned by the deadline, it is an emergency, and dozens, if not hundreds, of people will have to work under extreme conditions to find the "lost expedition." In general, there are strict laws in the mountains: discipline above all; never leave a comrade who has fallen off a rock face; and, if you have to, sacrifice your own life to save him.

Another rule is that the shortest soldier always leads the column and everyone else keeps in step with him. In the mountains, the pace of movement depends on stride length. If a column is led by a soldier who is six foot six, his stride length will be a lot for even for a person of average height to keep up with. A trailing short soldier will have to run, not walk. The soldier at the end of the line will soon begin to lag behind; which, by the way, is absolutely forbidden. **Remember**, the pace of movement must always be the same throughout the column.

Your right and left legs have different stride lengths, so in the dark or in a thick fog you will always move a little off the track. Therefore you have to adjust the direction of your movement frequently.

Mountain sickness can be a problem. There is only one way to deal with it—soldiers must train regularly at altitude over several years. Only a person who is physically fit, acclimated and trained to fight in the mountains need not fear this sickness.

Remember, do not leave any trash whatsoever behind. It is fairly easy for an experienced enemy to determine that your unit is there and how large it is from discarded items, chocolate wrappers and empty cans. It goes without saying that the enemy should not even guess that your group is in the mountains on a combat mission until the very last moment.

Get your soldiers accustomed to mountain wind, foul weather, sun and frost. Only then will your soldiers look on foul weather as an everyday occurrence. This is a factor for victory in battle. Foul weather is always an obstacle to your enemy as well.

You must train your soldiers so that they automatically select the best sites and firing positions. Small-arms firing is difficult in the mountains. It is difficult to fire accurately from one rock face at another, since you have to factor in meteorological conditions and your distance perception is distorted in the mountains.[2]

Remember that the best position for firing in the mountains is the prone position. You should always press as hard as possible against the rock face so that you are almost invisible. If, however, you decide to shoot while sitting or standing, a good shooter is sure to spot you and you can be sure he won't miss. The mountain units of foreign armies have spent years developing a well-worked-out technique for training Jägers.

What gear should you to take to the mountains? The law here is to take only the absolute necessities. Anything extra is going to be a drag on you both on the march and during the fight. I know from experience that when you first come into contact with the enemy, you will not be able to assume a firing position rapidly, much less change it. So all these jumar ascenders,[3] descent and arresting devices, pretty backpacks with hundreds of pockets, artificial warmers, and Chinese flashlights are simply going to end up discarded in the mountains.

Remember that your gear should depend on the mountain terrain and how you're your target is located. In a combat situation, you should take a climbing rope, a 20-25 foot reepschnur,[4] two-three carabiners,[5] a pair of climbing irons, glasses, and a pick. The carabiners are used for descent and for a lot more. The reepschnur can be used in lieu of jumar ascenders and descenders, and the pick can do everything: cut out steps in the ice, chip a stone off a rock face, arrest a fall from a rock face, or kill the enemy in hand-to-hand combat.

Be cautious about using imported gear. I know from experience that foreign carabiners such as the French- and German-manufactured Irbit have failed when someone falls from a rock face, if the gate has to bear the load. Remember that there is nothing better or more reliable than our Abalakovsky carabiner.

Footwear is important. In 1959, I hiked to the 22,906-foot Revolyutsiya Peak in ordinary felt boots and was comfortable, but when I went up the face a second time in shoes, my feet were cold. You need to choose climbing footwear very carefully, especially when going 16 20,000 feet or higher. Above all, remember to put Tricouni nails on your hiking footwear.[6] For example, some soldiers in Afghanistan's mountains did not put on Tricounis because they are fairly heavy and seem awkward at first. Many of these soldiers paid for this with their lives. And you will not, by the way, find better footwear than Tricouni.

With regards to cosmetics, remember that veteran mountain soldiers do not need creams, ointments or lip balm for protection against sunburn or the wind. It is only over time that wind and sun "harden" the skin and lips stop chapping. That is why I recommend that it is better to take a first-aid kit with a good variety of medications instead of cosmetics.[7]

Under no circumstances should you take alcohol with you on a mission to the mountains. The popular belief is that it warms you up in extreme cold. Take my word for it, this is an illusion. A person who has been drinking, and is tired after a long march to boot, will sleep like the dead in the cold and will not realize that he is freezing to death.

Remember to always have a sentry in the mountains. This is critical for ensuring that you accomplish your mission and return alive. Vigilance must be at the highest level. Anyone who fought in Afghanistan knows a lot of examples of weary soldiers assigned to sentry duty who fell asleep while on watch. The result was that the sentry, along with the unit, never did wake up: the Mujahideen slaughtered the sleeping soldiers like sheep, without firing a single shot.

And the main thing, in the mountains the commander should always be the most experienced and respected person. At least that is how it should be. Disaster is inevitable otherwise. The commander's professionalism as both a soldier and a mountaineer should therefore be head-and-shoulders above any rank-and-file soldier. This is the guarantee that all his orders will be carried out without fail.

The professionalism of the rank-and-file soldier is something that he gains only through years of hard combat training and systematic drills, as well as in marches and tactical exercises. In the final analysis, this is how all soldiers learn teamwork.

Remember that there is yet another rule in the mountains. Each soldier does what he is better at doing than anyone else. The soldier who knows about radio equipment is in charge of the radio and communications. Another soldier can always manage to light a fire in the mountains. He is in charge of campfires and oil stoves. The soldier who is best at stanching blood and bandaging a wounded comrade takes care of the wounded. The best shot has the sniper rifle and so on.

---

*Nazarov's tips seem like common sense, but they are common sense developed over forty years of mountaineering experience. The ancient Roman legionaires used to call themselves "Marius'mules" (*muli mariani*). The Emperor Gaius Marius (157-86 BC) initiated sweeping organizational reforms and greatly reduced the size of the logistics train by requiring each legionnaire to carry his armor, weapons, 15 days of rations (grain) and other gear. This onerous load weighed somewhere between 50 and 60 pounds. The normal days march was about 20 miles. The Roman Legion spent little time in the mountains. Today, the American infantryman goes into the mountains of Afghanistan carrying 85 pounds or more of light-weight gear. Afghans jokingly call the US Infantry the "heavy mules." After all these centuries, the soldier's load is still important—particularly in the mountains.*

Notes

1 Ikram Nazarov, "Malenkie secreti bol'shikh gor" [Small secrets of Great Mountains], *Armeyskiy sbornik* [*Army Digest*], May 2008, pages 35-36.
2 Small-arms fire invariably goes over the target when firing uphill or downhill. This can be corrected through training, but most armies only conduct marksmanship training on flat land.
3 A Jumar is a rope ascending/descending device that was first manufactured by the Swiss in the 1950s. It is a metal devise that fits on a rope and has a cam that allows the device to slide freely in one direction.
4 The Russians adopted a variety of German technical climbing words, among them reepschnur or rope-cord. This refers to an auxiliary cord five to six millimeters in diameter.
5 A carabiner (*karabiner* in German) is a metal loop (snap-link) with a spring-gate or screw-gate.
6 Tricouni nails (edge nails) are manufactured in Geneva, Switzerland, and are mounted on the sides of the soles of nailed climbing boots. They mount under the balls of the feet, toes and heel.
7 Sunscreen and Chapstick© are still essential until one develops a leather face.

5

Flanking Detachment In The Mountains: A Soviet Experience

LTC Lester W. Grau, US Army (ret.)

This article first appeared in the May-June 2010 *Infantry Magazine.*

Defending mountain terrain has its own challenges. Mountains offer good observation, but the terrain also blocks observation, particularly close up. Mountains offer good long-range fields of fire, but mountains are also full of dead space and concealment. Mountain defenses are not continuous, but are normally separate outposts and fighting positions which may be mutually supporting, but usually are not. They are often not even in the same plane. Mountain fighting positions are difficult to construct and maintain. Mountain fighting positions can be stockpiled with ammunition, but food and water quickly runs out at these positions and food and water are normally supplied in the villages and hamlets down in the mountain valleys and canyons. Consequently, in Afghanistan, the Mujahideen usually congregated in the valley except when they felt threatened. Some security was maintained at the fighting positions, but this was usually slack without indications or intelligence of enemy actions.

Attacking in the mountains has its own set of problems. First, the enemy holds the high ground and, if he has occupied the area for any time, he has had time to establish fighting positions and emplace long-range crew-served weapons such as mortars, heavy machine guns, recoilless rifles and even direct-lay artillery. He has had time to reinforce the defenses with mines and other obstacles. Entries into the mountain valley or canyon are limited and liable to interdiction by a skilled defender. Still, the irregular mountain terrain offers distinct advantages to the attacker. The enemy is seldom able to mass fires; the terrain offers numerous concealed attack approaches to defending positions. Enemy withdrawal will be by small groups and he will often be forced to abandon heavy weapons, ammunition stockpiles and wounded.

A Frontal Attack ... is not Recommended by Major V. A. Selivanov[1]

In April 1985, according to intelligence reports, there was a significant grouping of the armed opposition concentrated in the Mazlirud and Kakh Canyons. Their number was estimated at 1,200. Besides assault rifles, this group had 35-40 DShK heavy machine guns and up to 15 ZU anti-aircraft machine guns as well as mortars, recoilless rifles and rockets.[2] The main body of the enemy (400-600 men) was located in the village of Malakhairu. The general situation was complicated by the fact that earlier large-scale operations in the area showed that surprise was not possible and that, other than the casualties inflicted, the results were insignificant. The main body of the enemy, as a rule, managed to withdraw from the canyon before our troops arrived. The enemy had managed to establish a significant, well-constructed system of observation and early warning. Further, this region was particularly unsuited for air assaults and military vehicles could enter the canyon only on one road which ran through Mazilishakhr, Zagan, and Malakhairu (*see map on following page*).[3]

Considering the peculiarities of the region, during planning, our battalion commander determined to carry out the mission in the following fashion. We were briefed at an officers' call that at 1330 hours on the 9th of April, the *bronegruppa* of the main body would move from its base camp along the road from the south with the mission of closing in to the village of Mazlishakhr by 1800 hours.[4] My parachute company was ordered to support the departure and movement of the march column. Then, after the column had passed, my company was to return to garrison to carry out guard duty. The combat action, planned for the 10th and 11th of April, was to destroy the enemy grouping by moving the *bronnegruppa* to link up with its paratroopers who would move from the west into the valley.

However, at noon on the 7th of April, my regimental commander, without any witnesses, gave me an order for a totally different plan.[5] The new mission was a follows: at 1600 hours on 8 April, my parachute company would serve as a flanking detachment, and without attracting any attention, would secretly move to the region of Khakfakhai village in order to conduct a route reconnaissance for movement through the Afedshakh Pass. With the advent of darkness, I would leave my combat vehicles on the road, go on foot through the pass, enter into the Mazirrud valley and, by dawn, occupy positions on Khaduzan Mountain. At dawn on 9 April, I would be in position to adjust artillery fire and direct close air support with the goal of creating panic among the assembled Mujahideen and not allowing them to exit the canyon before the main body struck them. While carrying out this mission, very strict attention had to be paid to the secret movement and independent actions since it would be impossible to support my detachment in case we were discovered.

At 1800 hours on the 8th of April, my company was assembled in the designated area near Khakfakhai village. Soon, our observation disclosed that groups of armed men periodically moved on the path through the Afedshakh Pass. There was also a group of eight Mujahideen located in a guard post on that very same pass. Everything appeared calm in the target area. Everyone discussed it and agreed that the "ghosts"

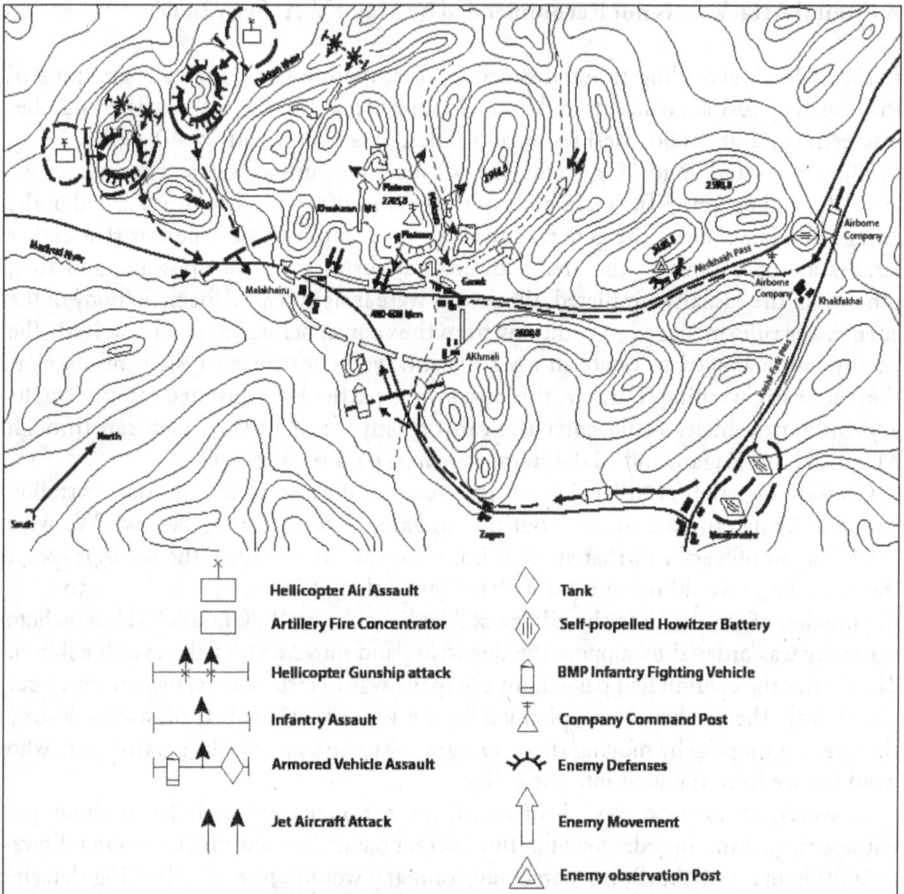

[Mujahideen] were not expecting our force. With the onset of darkness, we conducted a radio check with our armored vehicles and our higher headquarters and then I gave the order to begin movement.

We crossed through the pass by 2200 hours, bypassing the guard post to the south without drawing the attention of the Mujahideen. During night time, this post was a main [security] link. They were tied in with other posts using a complex system of varying signals. Even though these "stinkers" were one of the [security] links, we bypassed them literally in 15 minutes.

Just as we began to descend into the canyon, we lost radio contact with our company *bronnegruppa* and with our higher headquarters. By all rules of military science, I must immediately either restore communications or return.[6] But then our company's combat mission would not be accomplished. Therefore, I made a decision to continue to move and to restore communications once we reached high ground. I had to make this decision because I knew that if the Mujahideen discovered our company at night, we could only count on ourselves for help.

From the depths of the canyon, we heard the noise of night firing. We were convinced that the Mujahideen were not expecting the appearance of our force and had occupied their fighting positions to sort things out [conducted an alert]. At midnight, we entered the hamlet of Gusak. This is where I, the company commander, made a mistake that might have compromised all our measures for secrecy. Despite the ample number of night-vision devices that we had, we discovered too late that there were two Mujahideen patrols that we barely avoided running into. Taking necessary precautions, we lost about two hours before we exited the village. At 0300 hours, the company assembled at the foot of Khaduzan Mountain.

The heavens were clear and things were now visible thanks to the appearance of the moon. This additional lighting "worked" to the enemy's advantage and forced us to hurry up. The first platoon, commanded by Senior Lieutenant A. Mikenin, climbed the mountain. After an hour, he reported that his platoon had occupied positions on the heights. I placed two platoons, under the command of Senior Lieutenant V. Plotnikov, in an ambush directed against the hamlet of Gusak, and I followed the path of the first platoon to the top. We established our primary observation post on height 2705.0. I placed Lieutenant Mikenin's platoon on the northern slope and I placed the other two platoons on the southern slope. From these positions, they could also interdict paths on the eastern side and partially on the western side, while blocking enemy bands located in the hamlet of Gusak.

At 0500 hours, we were prepared to carry out our mission. At that time we were able to reestablish radio contact. Our under-strength paratrooper company had secretly crossed 12 kilometers of enemy-controlled territory, assembled in the rear of a strong enemy force and taken the commanding heights. It should be noted that the secret movement of my company was key to the success of the action and the complete lack of company casualities.

On the morning of 9 April, the *bronnegrupa* of the main body of the battalion inconspicuously passed through the village of Mazilishakhr and began to enter the canyon. Instantly, the signal (three individual shots) repeatedly rang from the mountain slopes from the north to the south and even the west. It announced the arrival of our force. Some 15 to 20 minutes after the signal, a band of 120-200 men emerged from the hamlet of Gusak and began to advance on my company. When the Mujahideen were close to our second and third platoon positions, my paratroopers opened up on them. The deadly fire caught the enemy completely by surprise and inflicted such heavy casualties on them, that they were unable to offer resistance. The "ghosts" panicked and ran back into the hamlet.

At this point, I should note that the sun was in our eyes and it was hard to find the enemy. After five minutes, the Mujahideen suddenly launched an attempt to break out of the canyon to the northeast. I called artillery fire on them. Again, after a half hour, they attempted to bypass the company to the north and up the southern slope of height 2946.6, but they came under the fire of Senior Lieutenant Mikenin's platoon.

From 1400-1500, we conducted two tactical air assaults [with the remaining two companies of paratroopers] on the western side of the canyon. The [*bronnegruppa* of the] main force, by this time, had already pushed through the hamlet of Gusak. The company's mission was over.

Thinking over our combat action as a flanking detachment in the mountains of Afghanistan, I have arrived at several conclusions. First, in order to conduct combat in similar circumstances, it is necessary to plan to assign an element of the combat formation to be a flanking detachment. It will be able to secretly enter the flank or rear of the enemy without engaging in combat with small subunits and guard posts. It will be able to prevent the enemy withdrawal, hit him with a surprise attack to destroy him and his capabilities in order to facilitate the successful mission accomplishment of the main force.

The experience of conducting such operations shows that most successful flanking detachments are company-sized. Use well-trained subunits and personnel in forming the detachment.

Second, the nature of mountainous terrain prevents small subunits from carrying heavy weapons and ammunition. At the same time, it is necessary to have sufficient fire power to conduct effective fires at various ranges. Besides our assault rifles and sniper rifles, my company carried one AGS-17 and a heavy machine gun.[7] In every squad, we had one AKM assault rifle with the under-barrel grenade launcher.[8] Our ammunition load was 600-700 rounds per assault rifle and 1200 rounds for the machine gun. Our special gear included some night-vision devices.

The successful actions of a flanking detachment in the mountains are dependent on close, well-planned coordination with the forces and resources of the senior commander. **Of primary importance is the support of artillery and aviation as well as agreement on the action with the main force before it departs to carry out its mission. Thus, in the course of combat it is impossible to fulfill the mission without reliable uninterrupted communications. Experience shows that in order to guarantee communications, it is better to establish retransmission stations or simply to use aircraft that are equipped for retransmission.**

In conclusion, selection of company-grade officers for assignment to a flanking detachment requires care. You know that they will be required to make independent mission decisions while separated from the main body on unfamiliar territory that is controlled by the enemy. This requires detailed planning and thorough preparation as well as a high degree of individual training. In part, it is absolutely necessary that the commander has concrete experience working with maps, can quickly detect objectives in the mountains, determine the necessary data for their destruction, precisely direct the fires of his subordinate units, and skillfully use all possibilities for secret and sudden actions.

In this article, the Soviets conducted the apparent main attack with their armored vehicles moving into the canyon on the only road. This attack included the personnel carriers of two paratroop companies, attached tanks, and an attached battalion of self-propelled artillery. There was little infantry in this attack. The two airborne companies that the armored vehicles belonged to conducted an air assault approximately six hours after the beginning of the supposed main attack to eliminate Mujahideen positions in the west and then to move into the canyon to link up with their vehicles. Helicopter gunships provided close air support to the air assault. The flanking detachment, which had inserted itself into the depths of the enemy position, was particularly useful in calling in artillery strikes and defeating an enemy advance and an enemy withdrawal attempt. Although the article provides no casualty figures, the flanking detachment suffered none while the Mujahideen did.

There are some interesting aspects to this attack. The flanking detachment left its armored personnel carriers (bronnegruppa) at the dismount point. It was responsible for its own security and served as a mobile reserve in the event that the flanking detachment got into trouble, but once the flanking detachment was deep into enemy territory, the reserve role was problematic. A self-propelled artillery battery joined the bronnegrupa for the attack the following morning. The Soviets had entered this war with a limited ability to conduct split-Fire Direction Center operations. By 1986, this capability was well developed. Soviet artillery usually accompanied the ground attack since it was more accurate than jet-aircraft in support in the mountains. Further, the artillery was often used in direct-lay against a stubborn enemy.

The flanking detachment was not part of the original battalion planning and was controlled by regiment. In the interests of operational security, the company's role and presence were not disclosed to the rest of the battalion. The company was basically on its own for at least 15 hours except on-call artillery and close air support. Both require communications and were iffy at night. Communications were a problem and remain such in the mountains despite modern technology. Mountains absorb radio waves and distort GPS signals.[9] Satellite communications are spotty and a saavy opponent could jam the GPS and satellite telephone receivers in the mountains. Ground retransmission units are hard to move, emplace and defend and retransmission aircraft are few and seem to go down for maintenance at critical points.

The infantryman's load remains a problem and in Afghanistan, where the bulk of engagements are beyond 500 meters, small-caliber supersonic bullets fired from short-barreled carbines are ineffective. The fight devolves to the machine gunners while the rest of the platoon tries to get involved. The Soviets issued the AK-74 with the thought that the infantryman could now carry more ammunition. Where possible, units such as this airborne unit, went back to the longer-range, medium weight 7.62 cartridge. Still, the weapons that the airborne carried did not give it an advantage over its opponent. The airborne company's position on commanding terrain did.

Going into the mountains is critical to gaining the initiative and bringing the fight to the enemy's sanctuary. Still, bulling into the mountain valleys and canyons without securing high ground or establishing a blocking force is futile. The lightly-equipped enemy will withdraw over familiar territory leaving his burdened attacker behind. Flanking detachments are an excellent way to shape the battlefield and hold the enemy in place for punishment.

Notes

1 Major V. A. Selivanov, "Nastupat' v lob...ne rekomendyetsya" [A frontal attack...Is not recommend], *Armeykiy Sbornik* (*Army Digest*), October 2009, 21-22.
2 The ZU is an anti-aircraft 23mm machine gun which comes in a single and dual-barrel model (the ZU-23-1 and ZU-23-2).
3 All place names in this article have been changed. The Russian Army has become hyper-security conscious to the point where even place names connected with historic events in official accounts are now being changed. Fortunately, this silliness does not extend to the veterans who are no longer on active service.
4 The *bronnegruppa* [armored group] consists of the tanks and personnel carriers of a unit which, after the infantry has dismounted the carriers, are used as a separate reserve. It is usually commanded by the unit's first deputy. In this case, the bronnegruppa has the BMPs of two companies (22 vehicles), probably at least two tank platoons (six tanks) and 12 self-propelled howitzers plus other vehicles supporting the mortar platoon, signal platoon and so on.
5 The Soviets always insured that orders and plans were witnessed and approved so that if things went wrong, the blame might be assigned or shifted. The regimental commander is undertaking a risky ploy, so he issues the order without any witnesses. Apparently, even the battalion commander does not know that one of his subordinate units has this mission. The company commander is on his own, if things go wrong.
6 The commander is already out there on his own without support and loses communication. By Soviet regulations, he is required to reestablish communications or abort the mission. This is a very risky decision.
7 The AGS-17 is a tripod-mounted 30mm automatic grenade launcher which fires rounds out to 1,700 meters. The machine gun was probably the 7.62mm PKM machine gun.
8 The GP-25 sub-barrel grenade launcher would fire a 40mm grenade out to 100 meters. These paratroopers preferred the older AKM Kalashnikov, with its 7.62mm round over the newly issued AK-74 with its 5.45mm round. Combat in the mountains is long-range and requires a subsonic medium-weight bullet. The AK-74 was designed for close combat and has less range and punch over distance.
9 In the mountains, GPS accuracy can slip from a five-meter diameter circle to a 500-meter diameter circle or more.

6

Cave Warfare

Ikram Karimov

This article first appeared in the February 2008 *Armeisky sbornik*

While reading several installments of "Komandirsky praktikum" ["Practicum for Commanders"] in *Armeisky sbornik* [*Army Digest*], I could not help but notice some inaccuracies and even errors in both obstacle negotiation and ready positions for shooting. But it is not just a matter of a few mistakes. Unfortunately, I did not see the most important thing in "Komandirsky praktikum": how to teach soldiers to fight in the mountains. I will therefore take the liberty of giving some examples of how a soldier should operate in so-called "cave warfare."

To begin with, let me remind you that 1999 and 2000 were very tense years for Uzbekistan.' Government troops were battling various separatist groups in the mountains. The telephone rang one evening and a Ministry of the Interior representative asked me to help prepare an operation to annihilate the separatists who were entrenched in the mountains and, while I was at it, train the civilian Ministry of the Interior personnel who were going to be the first to take part in such an endeavor.

I agreed and called in Ulugbek Pulatov, my old fellow mountaineer soldier and former paratrooper. We met at the staff office at the Sarisiye airfield and were quickly briefed in military fashion. As it happened, a battle was being fought in the upper reaches of the Kshut River, below the village of Tamarkhut, and in retreating the insurgents had taken up a defensive position in a cave that was hard to reach. It was on a cliff, by the confluence of a lateral tributary of the river, where mountain ridges with ravines formed a triangle in whose upper section was the cave with the insurgents in it— located at a height of 1,000 feet.

The rebels were ideally located, from a professional point of view: they had a view of the entire stretch from the confluence to the mountain range peaks, so it was virtually impossible to approach them unobserved. The government wanted to use a helicopter in the operation, but the mountain terrain ruled out maneuvering, and there was too great a risk that the helicopter would be shot down.

After carefully studying the mountain terrain, Pulatov and I saw the only possible way: approach the cave from above. A helicopter would land our group above the cave in the mountain where there was a more or less suitable landing area. We would then cautiously approach the spot where the sheer wall of the ravine in which the cave is located begins. In order to reach the precise spot, Ministry of the Interior personnel would have to make corrections to our movement by radio from points located opposite the wall, from where the cave entrance could be seen fairly well.

They watched the insurgents before setting the time for the attack. They selected early morning when the enemy would all be asleep. A group of six Spetsnaz soldiers were assigned to the group, three of whom were to make a sudden descent by rope and hang above the cave.

They formed three-man rope teams: one person would descend, another would belay, and the third would provide covering fire. It was important to gain time to hang opposite the cave entrance, which is why we decided that the Spetsnaz soldier who would descend first and hang above the entrance would throw a grenade and a smoke canister into the cave. And while the enemy was recovering, the rest of the Spetsnaz soldiers would descend and finish off the insurgents with assault rifles and grenades.

And after that it would be a matter of technique: the rest of the combat team would rush into the cave and finish off the resisting rebels. The entire group therefore numbered 30, of which 15 would directly participate in the attack while the rest would be in the support group.

Soldiers who already had combat experience were picked for these groups. They underwent mountain training exercises for several days and then conducted a whole series of battle drills in which the Spetsnaz soldiers trained to operate in conditions that were very close to those in which they would soon be going into action (*see figure*).

Having trained the groups, we expected that we would participate directly in annihilating the rebels that were entrenched in the mountains. However, the operation leader thanked us for our help and said that he did not have the right to risk civilian lives. We found out later that everything went exactly according to our scenario using only the Spetsnaz soldiers: the insurgents were caught unawares and, unable to offer serious resistance, were annihilated.

So the textbook that I am preparing now has a "Cave Warfare" chapter that offers practical advice on organizing a mountain battle using natural shelters, including caves.

Caves are well hidden and generally camouflaged by nature itself with shrubs or trees. They are excellent shelter. In guerilla warfare, it is convenient to use them in defense or for setting up firing ambushes. At one time, the Mujahideen in Afghanistan made firing emplacements out of mountain caves in particularly important areas along valleys and roads.

There was a case in which the Mujahideen sealed off a cave entrance with a large stone several meters in size which they would lift with powerful lifting jacks whenever

they wanted to shoot at a vehicle convoy of our "pourers" [fuel tankers] or at a motorized rifle company on the march. It was easy for them to monitor their enemy and shoot at us through the gap (embrasure) and then put the stone back in place after the attack. As a result, the mountain terrain on nearby slopes and ridges remained undisturbed. For a long time, both Afghan and Soviet soldiers who came under surprise fire could not understand how the Mujahideen managed to vanish so quickly in an unknown direction–until it dawned on them to set up a watch of the slope from which our units were fired on.

If an offensive is being planned, caves can be used as a supply storage base or a temporary infirmary, to keep prisoners in or a place for soldiers to rest in. Lastly, a cave can be turned into a convenient command post.

The Mujahideen in the Republic of Afghanistan made use of caves' tactical features: their distance from populated places, concealment, good camouflage in mountain terrain, and the difficulty finding and destroying them. In the notorious Tora Bora caves, they constructed wide corridors, multi-level galleries and halls, convenient passageways to firing positions with vehicular delivery of ammunition, and rooms for a hospital, recreation and headquarters.[2]

I remember a difficult situation that occured in the spurs of a mountain range. The enemy had occupied a cave. It was clear from the regularity of their radio traffic, that the insurgents were entrenched there. At night, they would go down to a nearby river for water, and in the daytime, they recharged solar batteries. Both the terrain and the cave's location were very similar to the one mentioned earlier. A small grassy slope with scrub went all the way up to the cave, and there was a small overhang— a rocky canopy that made it almost invisible.

We set up an ambush by the water. It was unsuccessful. Then in the morning, snipers opened fire on the lookouts and sentries from the ridges opposite, preventing the insurgents from detecting our soldiers in time. Our soldiers were, at that moment, rapidly descending by rope to the cave from above. Meanwhile, it was clear to everyone that as soon as just one of them appeared in front of the cave, they would be killed. But our soldiers were not only good mountaineers but experienced fighters as well. When they reached the cave canopy, they leaned forward as far as possible and threw grenades into the cave. The grenades were tethered to their right arms with cord.

When throwing a grenade on a cord, the distance and trajectory of its flight are measured visually. It takes systematic training to master this art. A grenade should not be thrown from the shoulder with a sweep of the hand but gently away from yourself, as if pushing it away, so that it swings like a pendulum into the cave. So when the grenade, like a pendulum, reaches the cave entrance and begins to swing inside, the end of the cord is released, thereby sending the grenade into free flight.

After the grenades exploded, the soldiers finished off the insurgents with assault rifles.

Experience shows that if the enemy continues to offer fierce resistance and shoot from deep within a cave, it is a good idea to use smoke canisters and teargas. A rocket-propelled anti-tank flame thrower (RPO) is particularly effective in this case. For example, in 1982, in Afghanistan, Captain S. Vlasenko successfully employed an RPO in a fight with Mujahideen who were entrenched in a cave near the town of Gogamunda (Jalalabad).

I conclude this article about techniques for annihilating an enemy that is entrenched in a cave by underscoring the following: the success of such actions depends first and foremost on the soldiers' physical, psychological and combat readiness, how well equipped they are with mountain gear, means of communication, and lightweight weapons, and the skill level of the support (covering fire) group. Developing the necessary skills requires daily training in mountains or in mountain training centers. It makes sense to include subjects such as the specific features of combat to annihilate the enemy in mountain caves in the mandatory training program.

Other major factors for success in a mountain battle are selection of the right time of day and weather conditions, as well as the psychological effect of our combat on the enemy that is defending the cave.

Notes

1. At this time, Uzbekistan, Tajikistan and Kyrgyzstan were involved in fighting a bloody civil war against the Independence Movement of Uzbekistan (IMU), which had trained and equipped in Taliban-controlled Afghanistan.
2. The author is probably confusing the caves at Zhawar with those at Tora Bora.

LARGE UNIT

7

The Campaign for the Caves: The Battles for Zhawar in the Soviet Afghan War

Lieutenant Colonel Lester W. Grau, US Army (ret.) and Ali Ahmad Jalali

This article first appeared in the September 2001 *Journal of Slavic Military Studies.*

The Soviets entered Afghanistan in late December 1979 to suppress a growing insurgency and to replace an out-of-control communist regime with another communist, pro-Moscow government. The Soviet government realized too late that they were then stuck in the middle of a civil war fighting guerrilla forces on some of the world's toughest terrain. The Soviets had planned to merely prop up the Democratic Republic of Afghanistan (DRA) and let the Afghan communist forces fight the Mujahideen guerrillas. Instead, the Soviets found their forces increasingly drawn into the battle — a nonlinear, unconventional battle which they were ill equipped or trained for.

The Soviet-Afghan War was primarily fought on the tactical level, but the strategic focus was a struggle by each side to strangle the others logistics lifeline. The Mujahideen and Soviets spent a great deal of time and energy defending their logistics network and attacking the other's. The Mujahideen targeted the Soviet lines of communication-the crucial road network over which the Soviet and DRA supplies had to move and ambushed convoys or cut off roads at critical mountain passes. The Soviet attack on the Mujahideen logistics was done in two phases. From 1980 until 1985, the Soviets sought to eliminate Mujahideen support in the countryside.

Early in the war, Mujahideen logistics requirements were fairly simple and primarily concerned with ammunition resupply and evacuating the wounded. The rural population willingly provided food and shelter to the Mujahideen, who were usually neighbors. The Soviets bombed rural villages and granaries, destroyed irrigation systems and crops, destroyed herds and launched sweeps through the countryside, conscripting young men and destroying the village infrastructure. The Soviet leadership believed Mao Tse Tung's dictum that the guerrilla lives in the population like a fish in water. They decided "to kill the fish by draining off the water."[1]

Afghanistan became a nation of refugees as more than seven million people left their farms and fled to neighboring Pakistan and Iran or to the cities of Afghanistan. The Mujahideen, who were used to living off the good will of the rural population, now had to transport rations as well as ammunition from Pakistan and Iran into Afghanistan.

The Mujahideen responded by establishing a series of supply depots, transfer depots and forward supply points logistics bases inside Afghanistan. These logistics facilities made it easier to provision the Mujahideen, but also had to be defended. The second Soviet phase was to find and destroy the Mujahideen logistics facilities. The odds are stacked against a defending guerrilla force since the attacker has the initiative, armored vehicles, air power and the bulk of artillery and fire power. The Mujahideen tried to offset this with a wise use of terrain and prepared defenses.[2]

Zhawar was a Mujahideen logistics transfer base in Paktia Province in the eastern part of Afghanistan. It was located four kilometers from the Pakistan border and 15 kilometers from the major Pakistani forward supply base at Miram Shah. Zhawar began as a Mujahideen training center and expanded into a major Mujahideen combat base for supply, training and staging. The base was located inside a canyon surrounded by Sodyaki Ghar and Moghulgi Ghar mountains. The canyon opens to the southeast facing Pakistan.

As the base expanded, Mujahideen used bulldozers and explosives to dig at least eleven major tunnels into the south-east facing ridge of Sodyaki Ghar Mountain. Some of these huge tunnels reached 500 meters and contained a hotel, a mosque, arms depots and repair shops, a garage, a medical point, a radio center and a kitchen. A gasoline generator provided power to the tunnels and the hotel's video player. This impressive base became a mandatory stop for visiting journalists, dignitaries and other "war tourists." Apparently, this construction effort also interfered with construction of fighting positions and field fortifications.

The Mujahideen "Zhawar Regiment," some 500 strong, was permanently based there. This regiment was primarily responsible for logistics support of the mobile groups fighting in the area and for supplying the Islamic Party (HIK) groups in other provinces of Afghanistan. Due to its primary logistic function, the regiment was not fully equipped for combat, but was a credible combat force. The regiment was responsible for local defense and for stopping infiltration of Khad and KGB agents between Afghanistan and Pakistan. They manned checkpoints along the road to screen identification papers. The regiment had a Soviet D30 122mm howitzer, two tanks (captured from the DRA post at Bori in 1983), some six-barrel Chinese-manufactured BM-12 multiple rocket launchers (MRL) and some machine guns and small arms. A Mujahideen air defense company defended Zhawar with five ZPU-1 and four ZPU-2 antiaircraft heavy machine guns. These 14.5-mm air defense machine guns were positioned on high ground around the base.[3]

Defense of the approaches to the base was the responsibility of Mujahideen groups from the National Islamic Front of Afghanistan (NIFA), the Islamic Revolutionary Movement (IRMA), and the two Islamic Party factions (HIH and HIK).[4] There were six major Mujahideen supply routes into Afghanistan.[5] Twenty percent of all

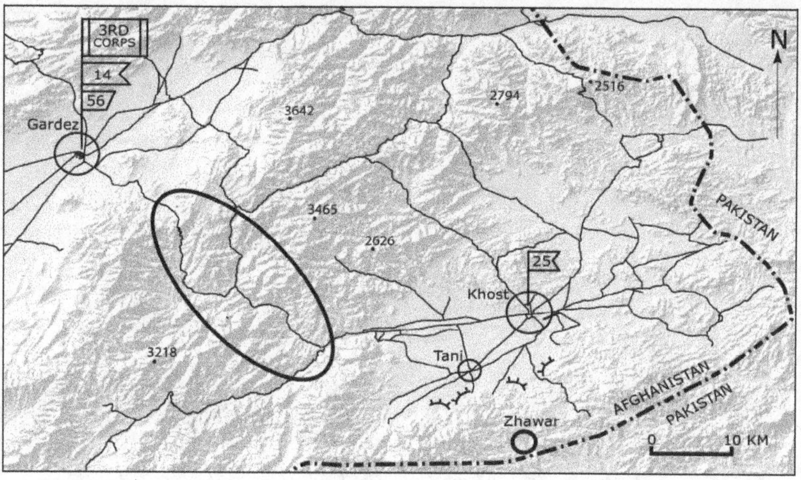

Zhawar and vivinity.

the Mujahideen supplies came through the Zhawar route.[6] The overall Mujahideen commander of Paktia Province, including Zhawar base, was Jalaluddin Haqani, who was a member of HIK.

Zhawar I

In September 1985, the DRA 12th Infantry Division from Gardez, with elements of the 37th and 38th Commando Brigades moved from Gardez circuitously through Jaji Maidan to Khost since the direct route through the Sata-Kandow Pass had been under Mujahideen control since 1981. This force joined elements of the 25th Infantry Division which was garrisoned in Khost. General Shahnawaz Tani[7] commanded this mixed force. The DRA military units had their full complement of weapons and equipment, but desertion, security details, and other duties kept their units chronically understrength. Since the DRA could not mobilize sufficient force from one regiment or division, they formed composite forces for these missions.[8]

Late one September afternoon, the DRA force began an infantry attack supported by heavy artillery fire and air strikes on Bori, which is northeast of Zhawar. Zhawar was not prepared for this attack since most of its major commanders, including Haqani, were on a pilgrimage to Mecca (the Haj). The DRA recaptured Bori and drove on to Zhawar. The Mujahideen reacted by positioning an 80-man group to block the ridge on the eastern slope of the Moghulgai mountains which form the eastern wall of the Zhawar base. The DRA force arrived at night and during the night fighting lost two APCs and four trucks. Eventually, the DRA became discouraged, withdrew and returned to Khost.[9] Mujahideen from the nomad Kochi tribe (led by Malang Kochi and Dadmir Kochi) and Gorbez Mujahideen, recaptured Bori.

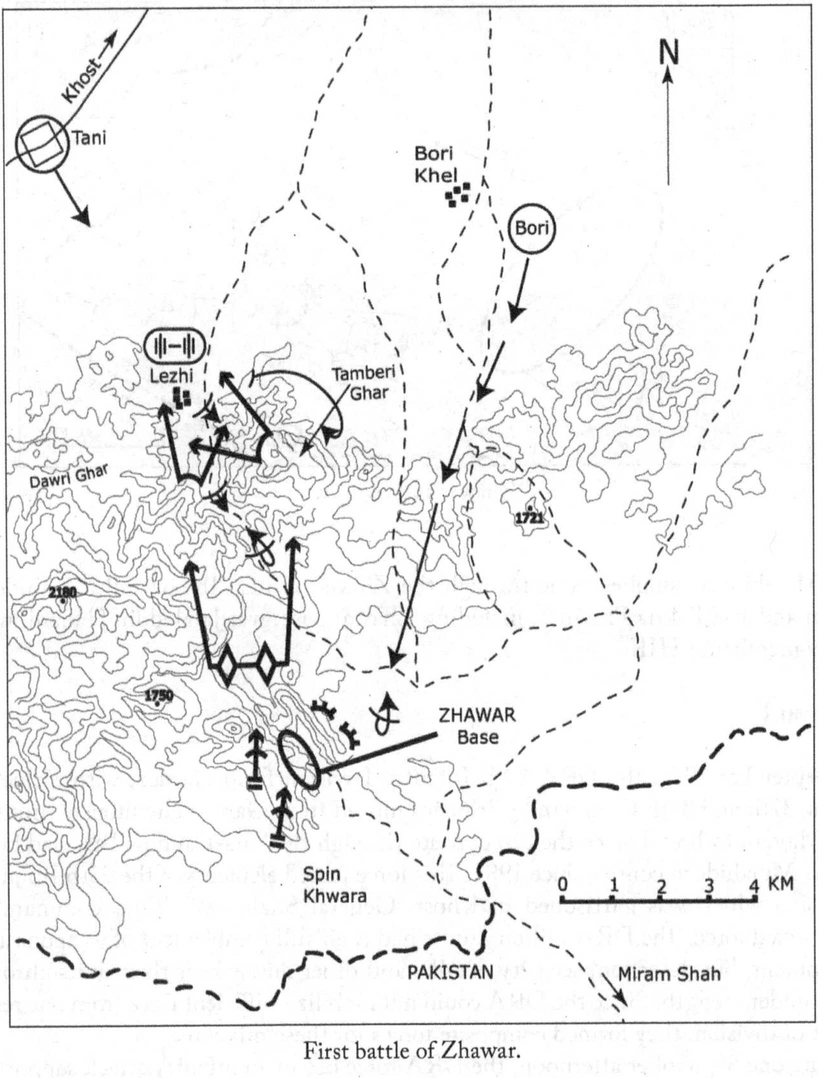

First battle of Zhawar.

The DRA then launched its next attempt from the town of Tani. They recaptured the town of Lezhi from the Mujahideen and killed Commander Mawlawi Ahmad Gul. The major Mujahideen commanders returned to Pakistan from the Haj on that day (September 4) and hurried north to Zhawar to take command. The Mujahideen from Lezhi retreated south while a 20-man Mujahideen force blocked the Manay Kandow pass. The pass is dominated by a high peak which is capped with a thick rock slab. Under the slab was a natural cave which the Mujahideen improved. The cave could accommodate the 20 Mujahideen during artillery and air strikes. The

Mujahideen also dug communications trenches so that they could quickly reoccupy their fighting positions once the firing stopped. The firing positions dominated the Tani Plain and were well positioned to stop any infantry attack.

The DRA repeatedly attacked the pass but could make no headway. The infantry would attack, meet withering Mujahideen fire and stop. Then massed air and artillery would pound the area. The infantry would again try to attack, but would again be stopped immediately. The procedure would then repeat itself, but the DRA made no headway during its ten-day attack. After ten days, the DRA called in heavy Soviet airstrikes which continuously hit the mountain top. The thick rock slab began to sway and rock. The Mujahideen were afraid that the rock slab might shift and crush their cave, so they finally withdrew. It was September 14, 1985.

As the Mujahideen fell back, the DRA established observation posts on high ground and started adjusting air and artillery strikes. This gave the tactical advantage to the DRA and their infantry moved through the pass. The Mujahideen rear guard desperately engaged the DRA infantry with machine gun fire and aircraft with ZGU machine guns. The DRA continued to advance and seized the high ground of Tor Kamar. Tor Kamar is within a kilometer of Zhawar base and well within the range of machine gun fire. The DRA thought that the Mujahideen did not have any heavy weapons and became careless and bunched their forces on the high ground. Two Mujahideen, Alam Jan and Muhammad Salim, were former tank commanders in the DRA. In the late afternoon, they moved their two tanks out of the caves and swung north into firing positions. They opened fire and their first rounds destroyed a DRA observation post as an artillery spotting scope and soldiers went flying. The Mujahideen tankers then traversed to the second observation post and destroyed it with their next rounds. Then they opened up on the other DRA soldiers.

The mauled DRA force fell back and maneuvered through the "bird droppings" saddle[10] to the east side of Tamberi Ghar. The Mujahideen countered with blocking positions which they held for five days. Haji Amanullah Khan and Ismail Khan played major roles in the fighting at this stage. The DRA Commander, General Tani moved his CP into the Many Kandow Pass and tried to reinvigorate the DRA assault, but the Mujahideen held. During the fighting, the Mujahideen shot down a helicopter, but lost a major commander-, Mawlawi Fathullah. Mujahideen reinforcements, including Commander Mawlawi Arsalah, arrived from Pakistan and as far away as Jalalabad and Urgun. The DRA were getting chronically low on men and supplies and, after 42 days of fighting, General Tani broke contact and conducted a night withdrawal.

Mujahideen casualties were 106 KIA and 321 WIA. DRA and Soviet losses were heavy, but their numbers are unknown since they evacuated their dead and wounded. Zhawar was a symbol of Mujahideen invincibility in the border area and the Soviets and DRA felt that they had to destroy this myth. The Mujahideen were convinced that Zhawar was impregnable and failed to take some basic security precautions. September-October and April-May are historically the best months in Afghanistan for campaigning, since the weather is reasonable and the roads are dry. August-September was also the time of the Haj and the senior leadership of the area all made this religious pilgrimage together. Consequently, the senior leadership was absent when the battle started and other Mujahideen commanders had to take command of the battle.

Field fortifications around Zhawar were neglected and incomplete. The excellent field fortifications at the mouth of the Manay Kandow Pass bought time to improve the other fortifications. A complacent attitude almost cost the Mujahideen their base. Only the unexpected appearance of Mujahideen armor at a crucial minute prevented a DRA victory. The Mujahideen were able to move men and supplies from Miram Shah in Pakistan throughout the battle. The DRA apparently made no attempt to impede access by deploying scatterable mines against the route.

Zhawar II

In February 1986, during the XXVII Congress of the Communist Party of the Soviet Union, General Secretary Gorbachev informed the delegates that the Soviet Government had worked out a plan with the Afghan Government to conduct a phased withdrawal of Soviet forces. The plan would immediately be put into effect after the political situation stabilized. The Soviet High Command issued orders to their forces to not get involved in direct combat when possible, but to emphasize security missions, guarding lines of communication and important installations. Simultaneously, they adopted additional measures to strengthen the DRA forces. The Soviets felt that the DRA should now take the leading combat role against the Mujahideen and urged the DRA to again attack Zhawar.[11]

The DRA Ministry of Defense decided to destroy Zhawar. General of the Army Varrenikov[12] approved the decision and the high command developed the plan for

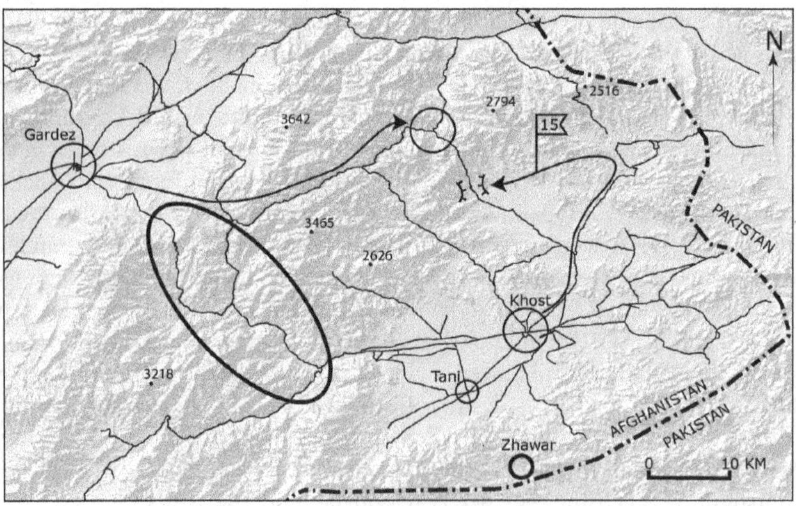

Prelude to Zhawar II.

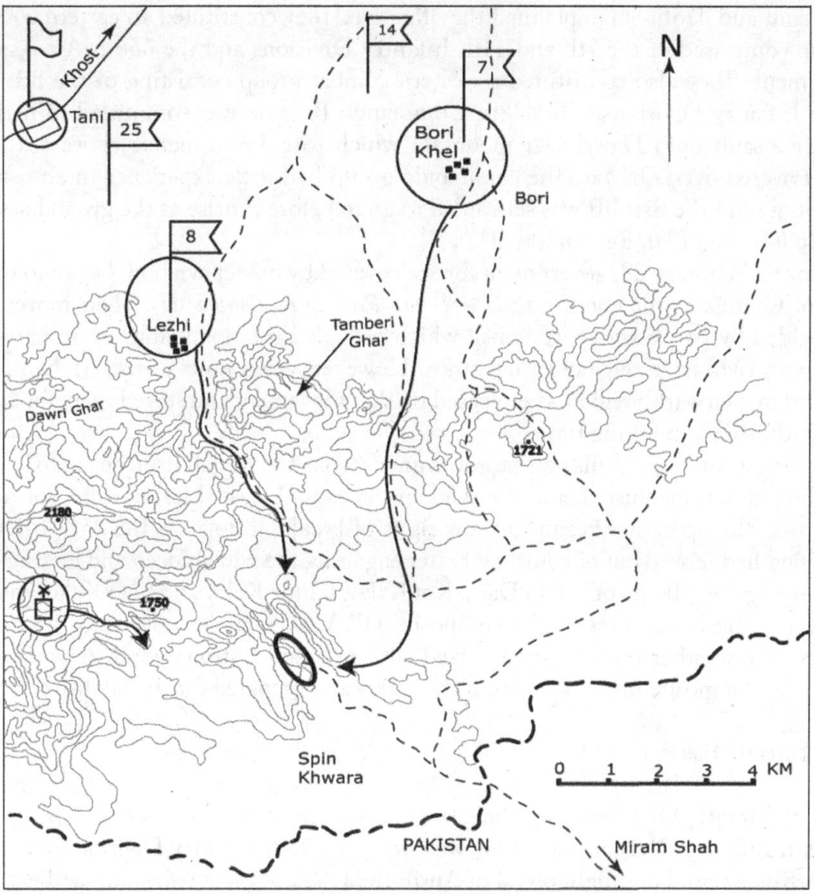

Plan for Zhawar II.

a combined operation.[13] The plan committed 54 under-strength DRA maneuver battalions plus DRA artillery and aviation to the assault.[14] The 7th Infantry Division (II Army Corps) from Kandahar andthe 8th Infantry Division (III Army Corps) from Kabul were committed to the fight. From III Army Corps' came the 14th Infantry Division from Gazni, the 25th Infantry Division from Khost, plus the 38th Commando Brigade and 666th Air Assault Regiment "Commando" also from Khost.

The commander of the III Afghan Army Corps, General-Major Mohammad Asef Delavar, was appointed the leader of the Afghan ground forces in this operation.[15] Deputy Minister of Defense General-Lieutenant Nabi Azimi was the overall commander of the Afghan Group of Forces. His adviser was Deputy to the Senior Military Adviser for Combat, General-Major V. G. Trofimenko.[16]

Azimi and Trofimenko planned the offensive. They constituted an eastern combat group comprised of the 7th and 14th Infantry Divisions and the 666th Air Assault Regiment. They also constituted a western combat group consisting of the 8th and 25th Infantry Divisions.[17] The 38th Commando Brigade was committed to make a an air assault onto Dawri Gar mountain which rose 3,600 meters above sea level and towered over Zhawar. The commando group had little experience in air assault missions, and the first lift was scheduled to go in before sunrise as the ground assault began (see map Plan for Zhawar II).

On the February 28, government forces, covered by Soviet aviation, began to move out of Gardez to the combat zone (*see map Prelude to Zhawar II*). Their movement was aided by two Soviet battalions[18] which occupied the dominant terrain between Kharzun (Mirazi Kalay) and Matwarkh. However, when units of the Afghan force arrived in Matwarkh region, they ceased further movement and stayed there for about a month, simply marking time.

Taking advantage of this passive government force, the Mujahideen began to launch shelling attacks against them. The Afghan forces took casualties, but did not move forward. The operation began to show signs of breaking down.[19] The 25th Infantry Division had moved out of Khost and after engaging the Mujahideen and looting and destroying the villages of Sekan Dara, Kot Kalay, Chine Kalay and Seto Kalay, moved to secure the Naray Pass so the composite DRA force could move into the Khost valley. The weather was wet snow mixed with rain and a strong wind. After several days, the composite force moved into the valley and prepared for the offensive.[20]

The Initial Disaster

The 7th Infantry Division was in the east with the 14th Infantry Division following and the 8th Infantry Division was in the west with the 25th Infantry Division following. Sometime around midnight on 2d of April, the DRA began a two-hour artillery and aviation preparation of the target area. Then six Mi-8 armed helicopter transport ships flew from Khost airfield to insert the initial assault group of the 38th Commando Brigade. The commandos landed without opposition, but the ground assault ran into immediate, heavy resistance from Mujahideen defending Dawri Gar Mountain. The ground advance was forced to halt.

The command post for the operation had moved from Khost to Tani and was in radio contact with the initial air assault group.[21] Despite all the Mujahideen fire, the air assault group commander reported that the firing was far away from his location. It was now 0300 hours in the early morning. The DRA artillery fired an illumination round on the northern slope of Dawri Gar mountain. "Do you see the round?" they asked the commandos. "Yes, we see it. It's about 15 kilometers from us" they replied. The DRA fired another illumination round five kilometers further away on the southern side of Dawri Gar mountain. "Do you see the round?" the command post again asked the commandos. "Yes, we see it. It's about ten kilometers from us" they replied. The commandos had landed some five kilometers inside Pakistan, beyond the base at Zhawar! The command post informed the commandos of the fact. The insertion group commander quietly answered "I understand. We will withdraw." But after an hour he reported that he was surrounded and locked in combat.[22]

The air assault was botched and someone, probably Generals Azimi and Trofimenko, made it much worse. They committed the rest of the brigade to combat — not onto the Dawri Gar mountain landing zone, which was well-populated with Mujahideen, but onto the open areas around Zhawar itself (see map Zhawar 2).

Mawlawi Nezamuddin Haqani was in the Zadran area when he saw approximately 20 transport helicopters flying over. He radioed the commanders at Zhawar and

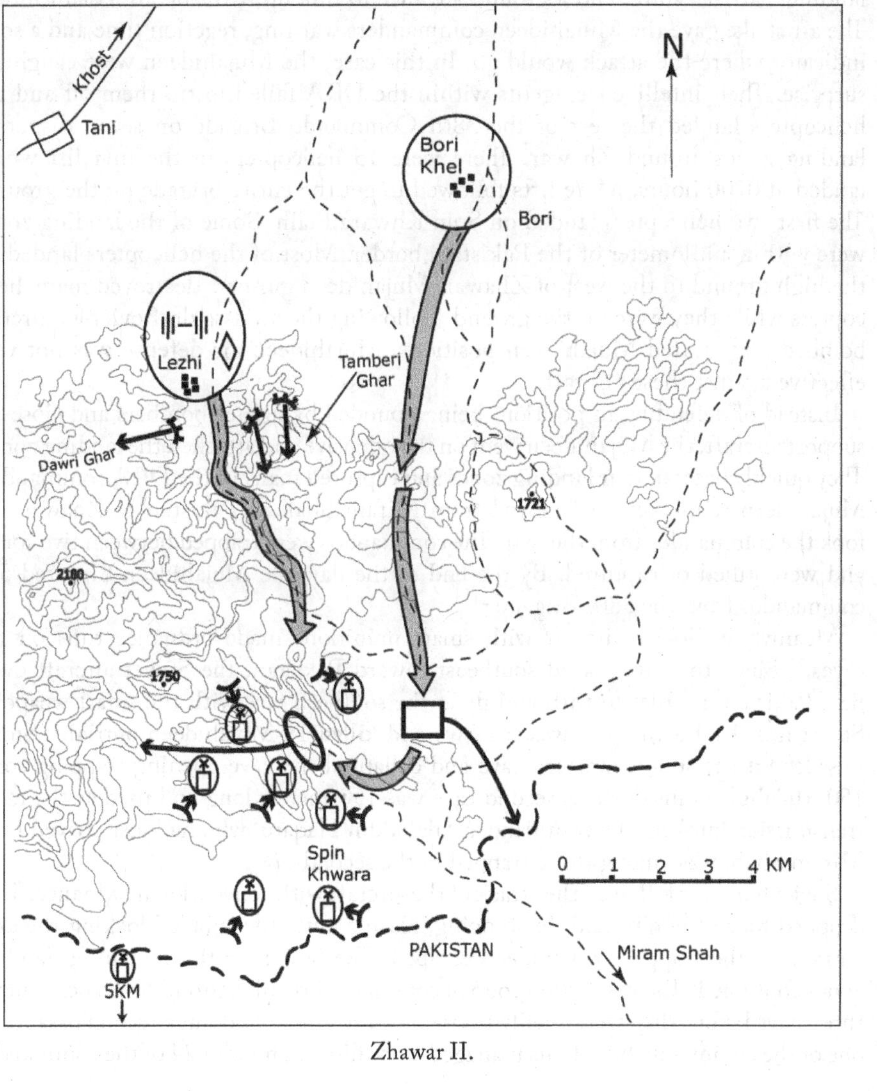

Zhawar II.

warned them. He expected that the helicopters would land at Lezhi or Darakai. After his radio message, he saw another group of helicopters, including some heavy transport helicopters, flying the same direction. These were escorted by jet fighters. He again radioed this information to Zhawar. Zhawar had 700-800 Mujahideen combatants, plus air defense forces, at the time. However Jalaluddin Haqani, the Zhawar Commander, was at Miram Shah. Mawlawi Haqani also radioed this information separately to Jalaluddin Haqani, who set out immediately for Zhawar.[23]

The usual Soviet/DRA pattern for an attack on a Mujahideen base was to pound the area heavily with air strikes and then follow the air strikes with air assault landings, artillery fire, and a ground advance to link up with the air assault forces. The air strike gave the Mujahideen commanders warning, reaction time and a solid indicator where the attack would go. In this case, the Mujahideen were caught by surprise. Their intelligence agents within the DRA failed to tip them off and the helicopters landed the rest of the 38th Commando Brigade on seven dispersed landing zones around Zhawar. There were 15 helicopters in the first lift which landed at 0700 hours. More lifts followed to get the entire brigade on the ground. The first two helicopters landed on Spin Khwara Plain. Some of the landing zones were within a kilometer of the Pakistani border. Most of the helicopters landed on the high ground to the west of Zhawar. Mujahideen gunners destroyed many helicopters while they were on the ground. Following the air assault, Soviet jet aircraft bombed and strafed Mujahideen positions. Mujahideen air defense was not very effective against these aircraft.

Instead of defending in positions being pounded by fighter-bombers and close-air support aircraft, the Mujahideen went on the offensive and attacked the landing zones. They quickly overran four landing zones and captured many of the DRA commandos. Mujahideen reinforcements moved from Miram Shah in Pakistan to Zhawar and took the commandos from the rear. The commandos were trapped between two forces and were killed or captured. By the end of the day, the Mujahideen captured 530 commandos from the 38th Brigade.[24]

Meanwhile, Soviet aircraft with smart munitions made ordnance runs on the caves.[25] Since the caves faced southeast toward Pakistan, the Soviet aircraft overflew Pakistan in order to turn and fly at the southern face with the smart weapons. Smart missiles hit the first western cave and killed 18 Mujahideen outright. Smart missiles hit the second western cave and collapsed the cave opening trapping some 150 Mujahideen inside. This second cave was 150-meters long and used as the radio transmission bunker. The commander, Jalaluddin Haqani, who had just arrived from Miram Shah, was among those trapped in the second cave.[26]

Soviet bombers followed the attack of the aircraft with conventional ordnance. They dropped tons of bombs and, in so doing, blasted away the rubble blocking the cave entrances. The trapped Mujahideen escaped. The battle for the remaining landing zones continued. There was one group of commandos on high ground who held out for three days before they were finally overrun.[27] The chief of counter-reconnaissance in one of the commando battalions managed to exifiltrate and lead 24 of the commandos to the safety of their own forces. It took eight days. Of the 32 helicopters assigned to the mission, only eight survived.[28]

The 7th and 14th Infantry Divisions were in the corps first echelon. They tried to link up with the already destroyed air assault-force. In the course of three days, they shot up their entire supply of ammunition and lost control. At the end of April 9, these divisions pulled back to their start points. The 25th Infantry Division, located in the second echelon, covered the western flank, the artillery positions and the corps rear area.[29] The 14th Infantry Division covered the eastern flank. The DRA had regarrisoned Lezhi since Zhawar One and continued to fight for the possession of the Manay Kandow Pass for some ten days following the air landing. The Mujahideen attacked the DRA LOCs and the airfield at Khost while the Mujahideen holding the Manay Kandow checked their advance.[30]

General Varrenikov sent a message to the Soviet Minister of Defense in which he criticized the leadership of the 7th, 8th and 14th Infantry Divisions and the III Corps Commander. He presented various alibis (weather, length of campaign, poor intelligence) and outlined his plan to reinforce the effort with three DRA regiments, a DRA spetsnaz battalion, and six Soviet battalions. He noted General Azimi's replacement as operations commander and requested time to resupply and prepare the force to resume the offensive. General Sokolov, the Minister of Defense, responded with a stern reply and gave Varrenikov twelve days to prepare for resumption of the operation (see Annex 1).

Once Again

The Soviet and DRA military leadership were in damage control. DRA reinforcements included a regiment each from the 11th and 18th Infantry Divisions and the 21st Mechanized Infantry Brigade as well as the 203d Separate Spetsnaz Battalion (special forces) and 37th Commando Brigade.[31] Following urgent requests from the leadership of the DRA, Varrenikov authorized five battalions of Soviet forces which were sent to Khost and Tani between 5 and 9 April.[32] Soviet Forward Air controllers were assigned to work with Afghan Forward Air Controllers in the infantry divisions and the reinforcing Soviet unit commanders were assigned to work with the Afghan division commanders.[33]

General-Lieutenant Azimi flew off to Kabul on "important business." From Kabul, he issued orders to arrest the helicopter regiment commander, but the commander was hiding. The helicopter pilots who landed the commandos in Pakistan said that the commander of the commandos had ordered them to land where they did.[34] General-Lieutenant Gafur from the DRA Operations Section of the General Staff replaced General-Lieutenant Azimi. The Chief of Staff of the 40th Army, General-Major Yu. P. Grekov took command of the five Soviet battalions. General-Lieutenant V. P. Grishin (Operations Group of the Ministry of Defense, USSR in Afghanistan) assumed overall coordination of all the forces. They reworked the operations plan while the force was refitted. The total DRA/Soviet force now exceeded 6,600 men.

Then Varrenikov himself arrived at the battlefield. DRA President Karmal had requested that General of the Army V. I. Varrenikov take over as overall commander of the operation.[35]

While this refitting, restructuring, and replanning were going on, the communists kept the pressure on the Mujahideen with air strikes and artillery. In first battle for Zhawar, DRA/Soviet artillery and air strikes stopped at night, but this time they were conducted around the clock. At night, they dropped aerial flares for illumination. This heavy fire support continued for 12 days.[36] The tempo of the air and artillery increased on the morning of April 17. Said Colonel S. Korennoy:

> Soviet pilots showed a miraculous mastery and heroism. Many of us saw how the aircraft of LTC A. Rutskoy, the commander of the aviation regiment, was shot down. ... His aircraft seemed to draw all the air defense fire. He made four or five runs on the base and then we saw his plane dart and he then flashed from the mountain ridge into the valley. We felt, from the convulsive jerks, that the pilot was attempting to start his engine, but alas. The bang of the ejection seat rang out. The aircraft flew straight and level for a few seconds and then the nose dipped and the aircraft slammed into the ground and exploded somewhere in the vicinity of Barankhel. A BTR from the operations group of the 40th Army went and picked up the pilot.[37]

Pakistan was clearly concerned with the major battle raging on her border. The Mujahideen lacked effective air defense against helicopter gunships, and the strafing and bombing attacks of high-performance aircraft. The Mujahideen had some British Blowpipe shoulder-fired air defense missiles, but they were not effective. Pakistan sent some officers into Zhawar during the fighting to take out attacking aircraft with the Blowpipe and show the Mujahideen how it was done. After climbing a mountain and firing thirteen Blowpipe missiles to no avail, a Pakistani captain and his NCO were severely wounded by the attacking aircraft.[38]

The renewed ground attack began on the morning of April 17. The 25th Infantry Division led the assault in the west and the 14th Infantry Division led the assault in the east. In order to deceive the Mujahideen and divert their forces, the eastern group began its attack at 0630 and the western group began at 1030.[39] The DRA 25th Infantry Division was concentrated at Lezhi. The Mujahideen had fortified the Dawri Gar Mountain and could cover the majority of the slopes with accurate fire. Multiple attacks on the mountain failed. When the artillery fire preparation would start, the Mujahideen would shelter in caves and when the preparation ceased, they reoccupied their firing positions and repulsed the attack. General-Major Asef, the DRA 25th Infantry Division Commander and Lieutenant Colonel Mikhail Karaev, his adviser, observed this. During the night, they silently moved one of their regiments toward the summit and, at dawn, launched an attack on the Mujahideen without any artillery preparation.

The Mujahideen did not expect this and faltered. The regiment captured the summit in a matter of minutes.[40] The Mujahideen fell back from the Lezhi area into the higher mountains and slowly the DRA/Soviet force moved through the Manay Kandow Pass.

At the same time, the DRA/Soviet force launched a flanking column from the Lezhi area that moved to the east. This column moved toward Moghulgai Mountain on the east flank of Zhawar. There, a regiment of HIH Mujahideen waited in defense. However, as the DRA column neared, the HIH regiment withdrew without a fight. At the same time, Jalaluddin Haqani was wounded by attacking aircraft. He had head and facial wounds, but rumors spread among the Mujahideen that Haqani was dead. The Mujahideen evacuated Zhawar and moved high into surrounding mountains as the two ground columns closed onto Zhawar.[41]

The Mujahideen were unable to evacuate most of the stores from Zhawar. They pulled out the two T-55 tanks and fought the advancing column for awhile before abandoning the tanks in the foothills. Lieutenant Colonel Kulenin, the adviser to the commander of the DRA 21st Mechanized Brigade and his political deputy were killed by a T-55 round.[42] The Soviet and DRA forces entered Zhawar. It was noon on 1April 19, 1986.

A Short Stay

As Viktor Kutsenko recalls: [43]

> After a narrow passage of mountain road, it opened up into a wide canyon of 150 meters, whose sides stretched upwards for two kilometers. Caves were carved into the rock face of the side facing Pakistan. The caves were up to 10 meters long, four meters wide and three meters tall. The walls were faced with brick. The cave entrances were covered with powerful iron doors which were painted in bright colors. There were 41 caves in all. All had electricity. Behind a fence there was a mosque with a beautiful brick entrance and a hospital with new medical equipment manufactured in the United States. They even had an ultrasound machine which we moved to the Khost hospital. There was nickel-plated furniture including adjustable beds. There was a library with English-language and Farsi-language books. There was a bakery and by the entrance was a stack of fresh nan.[44]
>
> In the storage area, there were metal shelving units where boxes of arms and ammunition were neatly stacked. Further on, there was a storage cave for mines. There was every kind of mine imaginable: antitank, antivehicular, and antipersonnel mines from Italy, France, the Netherlands, and Germany. Hand grenade and artillery simulators were stored separately. The demolition explosives of various types and detonators were stored in a separate cave. In the very furthest part of the base were repair and maintenance bays complete with grease pits. There was a T-34 tank in one of them. The tank was serviced, fueled and had new batteries. It started right up and drove out of the service bay. Above the storage caves was a beautiful building marked "Hotel." There was overstuffed furniture

> inside and the floors were covered with carpets. How many of our aircraft had worked this site over and the hotel and caves were still intact.

The Afghan soldiers began to loot the base. Even the two-meter high brick facing wall was pulled down and hauled back to the 25th Infantry Division at Khost. The DRA had no intention of staying in Zhawar long enough for the Mujahideen to organize a counterattack. The Mujahideen were moving MRL up to the Pakistan border to fire on the communist forces.

Colonel Kutsenko was in charge of destroying Zhawar and had four hours to do so. He split up the detonation of the caves and buildings between the sappers of the 45th Engineer Regiment of the 40th Army and Afghan sappers. He knew that he could not destroy the caves in the available time. Above the caves was a 30-meter thick layer of rock. If they could drill a one-to-two meter shaft into the cave ceiling, they could have crammed that full of explosives and caused a collapse, but there was not time to do that before the troops had to leave. So the sappers stacked about 200 antitank mines in the primary caves and rigged them for simultaneous electric detonation. Even if they had laid ten times more explosives, it would not have made any difference since the force of the explosion would follow the path of least resistance and the caves would channel the force out the caves' mouths.[45] Said Kutsenko:

> And the moment finally arrived. The caves ... shot out their contents. After the dust settled, all of the canyon was filled with clumps of earth, shattered bricks and stones. The caves were swept clean, but were somewhat larger and their entries were partially clogged by rock slides from above. The gates were torn pieces of iron laying at the foot of the opposite canyon wall.[46]

The combat soldiers were withdrawing as the sappers remained behind to mine the base. The work was hard and complicated by the lack of time. The sappers had to depart before nightfall. At 1700 hours, the command was given for the remaining force to leave and head for Tani. The Mujahideen were quick to fall on the heels of a retiring foe. Anyone who would fall behind or stop would be in serious trouble. Rockets fired from across the Pakistan border were landing near the sappers and these rounds were becoming more precise. It was time for the sappers to join the exodus. Kutsenko gave the command to depart on his radio. The Afghan sappers immediately quit working and boarded their armored vehicles. The Soviet commander of the sappers from the 45th Engineer Regiment answered "Right away." His "right away" lasted 15 minutes. Kutsenko again called him and ordered that they cease work and depart. Their commander again answered "Right away." Kutsenko then I told him "You may stay here for an hour, but your soldiers need to quickly join the convoy. The Mujahideen are here and we are leaving." This time, the Soviet sappers quit work and immediately boarded their vehicles. Kutsenko insured that everyone was on board and the trail party left. Kutsenko sat in the captured tank and returned to Tani on it.[47]

After 57 days of campaigning, the DRA held Zhawar for only five hours. In addition to the standard mines and booby traps, the communist forces planted seismic-detonated mines and sprinkled aerial-delivered butterfly bombs over the area. The Mujahideen returned to Zhawar on the following day. The first Mujahideen to enter the area were killed by seismic mines. The Mujahideen withdrew and fired mortars, BM12 and machine guns into the area to set off the seismic mines. Then they began the slow process of finding the rest of the mines manually. The Mujahideen pushed forward from Zhawar to retake Lezhi and other areas. Since the DRA was only in Zhawar for five hours, the DRA did not manage to destroy the caves, but collapsed some entrances. Weapons that were stored in some of the caves were still intact and useable.[48]

Mujahideen casualties were 281 KIA and 363 WIA. DRA and Soviet losses are unknown, but the Mujahideen reportedly destroyed 24 helicopters, shot down two jets and captured 530 personnel of the 38th Commando Brigade. The Mujahideen held a field tribunal. Yunis Khalis and other Mujahideen commanders were the judges. They tried and executed Colonel Qalandar Shah, the commander of the 38th Commando Brigade and another colonel who landed with the brigade to adjust artillery fire. There were 78 other officers among the prisoners. They were given a chance to confess to their crimes from different battles and then all the officers were executed. All the soldiers were given amnesty since they were conscripts who were forced to fight. The amnestied soldiers were asked to perform two years of labor service in exchange for the amnesty. They did their service in logistics, were "reeducated" and released after two years.

Aftermath

The withdrawal of the HIH regiment, coupled with the rumors of Haqani's death, greatly aided the DRA victory. Haqani's loss, besides affecting Mujahideen morale, cost the Mujahideen what little command and control they had left at this juncture of the battle. The DRA failed to throw a blocking force on the Miram Shah road, although they knew that Mujahideen reinforcements were moving along this route. As a minimum, they could have employed scatterable mines on the road, but they left the route open.

Clearly there were intelligence failures on both sides. The DRA and Soviets had ample opportunity to collect information on Zhawar, but failed to determine the basic outline of the Mujahideen defenses and their manning. The DRA and Soviets never detected the presence of the Mujahideen regiment that slipped away at the crucial juncture and allowed their victory. The Mujahideen, on the other hand, should have been aware that Zhawar was at risk since the DRA had been moving forces toward the Khost valley for some forty days. Yet, the leadership of Zhawar was out of country during the lengthy build-up. The usual Mujahideen sources failed to tip them off as to the start time of the assault or the air assault. And, although the pathfinder commando element that landed in Pakistan was overrun by Mujahideen, these Mujahideen did not get the word to Zhawar. The Zhawar defenders were surprised by the air assault of the commando main body.

The DRA and Soviet reluctance to hold Zhawar for any length of time in order to do a thorough job of destroying the base is a strong testament to the ability of the Mujahideen to threaten their lines of communication. The commanders had no desire to risk being trapped in Zhawar and having to mount yet another operation to fight their way in and out of Zhawar. Their reserves were committed and the danger was real.

The DRA celebrated the fall of Zhawar with parades and medals as a major victory. But Zhawar was back in full operation within weeks of the attack. Having been trapped in the caves, the Mujahideen learned to make connecting tunnels between caves. They reopened the caves and built connecting tunnels. The caves were improved and lengthened to 400-500 meters long.[49] In retrospect, the battles of Zhawar seem to have been exercises in futility, but at the time, they were considered tests of whether the DRA could stand up to the Mujahideen after the Soviet withdrawal.

Annex 1: Correspondence between General Varrenikov and General Sokolov
SECRET (URGENT)

USSR Ministry of Defense
To Marshal of the Soviet Union Comrade S. L. Sokolov

I report.
Preliminary results from the combat conducted by the Afghan Army in the region of Khost may be evaluated as unsatisfactory, although the Mujahideen suffered significant personnel and weapons losses.

Reasons

1. The forces and aviation assigned to the operation were poorly trained and the personnel had poor morale and fighting spirit. The combat potential of the divisions was weak and their potential was very limited. Further, the commanders of the 7th, 8th, and 14th Infantry Divisions and, especially, the commander of the III Corps were completely unprepared to lead their forces.

2. At the start of the operation, there was an incomplete intelligence picture and a wrong estimate of enemy strength. Enemy strength was much greater than determined. The close proximity of Pakistan allowed the enemy complete maneuverability and the unimpeded capability to replace personnel and weapons losses. Further, the Mujahideen were able to use strong fire support from the territory of Pakistan. This was particularly true against our forces moving along the border.

Young, well-trained, steadfast Pakistanis participated directly in the battle in the base region of Zhawar and Miram Shah (south of Khost).

3. This was an unfortunate time of year to conduct combat. A more advantageous time would have been in January to the start of February or in April. The end of February and March are characterized by an abundance of rain mixed with snow (particularly this year). This forced the troops to move through the thawed mud along the existing road under enemy fire.

4. The difficult weather, length of the combat (30 days) and the "uncomfortable" column formation forced on the troops in their advance to the international border exhausted the troops and led to low morale. The short preparation time for combat in Khost from 31 March to 4 April did not revitalize the forces.

5. During the course of combat, the leadership tolerated nonobjective estimates of the situation, false situation reports and false combat reports. As a result, they were unable to provide the necessary picture of the situation.

6. There were mistakes in the conduct of the assault landing. The planning was conducted correctly. However, the practical application was organized unsatisfactorily. Because of this, the assault force was scattered over LZs located four to 20 kilometers from their planned LZs and this allowed the Mujahideen to defeat the air assault in the course of a day. The preparation of the helicopter crew navigators was unsatisfactory and they were poorly oriented on the terrain during the landing. They only had bearings and times of flight. These were the main reasons for the errors.

Measures that were taken

1. Additional massive air and artillery strikes were planned and conducted on strong points and pin-pointed enemy firing points. The "Shleyf" round was employed (on the approaches to the base, strong points equipped with cement structures, and armored cupolas of dug-in tanks).

2. Reinforcing units were moved into the combat zone. These included the 50th Infantry Regiment from the 18th Infantry Division, the 81st Infantry Regiment from the 11th Infantry Division, an infantry regiment from the 21st Mechanized Infantry Brigade and a battalion of spetsnaz.

Soviet forces were moved to support the Afghan forces. These included two battalions from a separate air assault brigade and two battalions from a separate parachute regiment.

3. All types of intelligence collection were reinforced, particularly that directed against the Zhadran tribe.

4. The leadership of the Afghan forces was reinforced. With this goal, General-

Lieutenant Azimi (who was ill) was replaced by General-Lieutenant Gafur to direct the battle....

The Army Chief of Staff directed the 40th Army units. Overall coordination of all the forces was provided by General-Lieutenant V. P. Grishin (Operations Group of the Ministry of Defense, USSR in Afghanistan).

Preventive Measures (lessons learned)

1. It is necessary to take radical measures to replenish the personnel in the Afghan Army (particularly in the 1st and III Army Corps). Right now, the line strength of the combat divisions is 200-300 men. ... [Desertion, combat loses and an inefficient conscript system bled divisions down to mini-battalion size.] ...

5. Refrain from conducting large-scale independent actions by the Afghan forces, particularly those that last a long time or are conducted far from their garrisons. Combat must not exceed 10-12 days.

6. The forces must not conduct actions against Mujahideen bases located close to the border or in a region where it is impossible to isolate the enemy from the arrival of his reserves. These regions should be subjected to massive aviation strikes, dropping powerful bombs, scatterable mines, etc.

Further actions

... Consider that the result of the conduct of military actions will convey a significant military-political ideal, it is expedient to increase the aviation strikes to the maximum, to destroy enemy points and simultaneously take measures to train the Afghan forces for decisive battle....

Varrenikov, April 1986[50]

The Minister of Defense of the USSR replied with the following order to the Operations Group of the USSR Ministry of Defense in Afghanistan:

SECRET

First Deputy Chief of the General Staff
General of the Army Comrade V. I. Varrenikov

Chief Military Adviser in the DRA
General-Colonel Comrade V. A. Vostrov

The Chief Military Adviser in the DRA and his operatives were mistaken in their estimate of the size of the enemy groupings and their potential in Paktia Province. During the preparation for combat, they did not consider the changes which had occurred since 1985 in that region. The Mujahideen have received new weapons and equipment and learned from the experience of the 1985 strike in that region. The timing of the operation was a mistake. In the course of the operation, neither the new Chief Military Adviser to the DRA, General-Colonel V. A. Vostrov, nor the Operational Group of the USSR Ministry of Defense took the necessary measures in order to correct the errors in a timely manner.

I demand:

1. The conduct of a comprehensive estimate of the correlation of forces assembled in the region of Khost. Improve intelligence activities against the enemy. Conduct air strikes based on targets identified by intelligence and cease area bombing.

2. Organize troop control. Subordinate the reinforcing units brought from the center of the country to the appropriate division commanders before accepting combat again.

3. If necessary, delay your offensive actions for several days. Ready the units and subunits of the Afghan forces for the upcoming combat. Replenish their ammunition and material stocks and clarify their combat missions. Organize coordination with aviation.

4. Your plan for the conduct of future combat will be present for confirmation by 17 April.

Sokolov

312/1/07 sh
April 1986[51]

Ali Ahmad Jalali, former Interior Minister of Afghanistan (January 2003- September 2005), is currently serving as a Distinguished Professor at the Near East South Asia Center for Strategic Studies (NESA) at the National Defense University in Washington D.C. He served as the Director of Afghanistan National Radio Network Initiative and Chief of the Pashto and Persian Services at the Voice of America in Washington D.C. He covered the war in Afghanistan (1982-1993) and extensively traveled in the former Soviet Central Asian republics. A former officer in the Afghan Army, Colonel Jalali served as a top military planner with the Afghan Resistance following the Soviet in- vasion of Afghanistan.

He graduated from high command and staff colleges in Afghanistan, the United Kingdom, the United States, and the Soviet Union. A published writer in three languages (English, Pashto, Dari/Farsi), Ali A. Jalali is the author of numerous books and articles on political, military and security issues in Afghanistan, Iran and Central Asia.

Notes

1 Claude Malhauret, Afghan Alternative Seminar, Monterey, California, November 1993.
2 Ali Ahmad Jalali and Lester W. Grau, *The Other Side of the Mountain: Mujahideen Tactics in the Soviet-Afghan War*, to be published in 1998, xvii and Chapter 11, page 1.
3 Interviews with participants Lieutenant Omar and Mawlawi Nezamuddin Haqani in Jalali and Grau, chapter 11.
4 Mohammad Yousaf and Mark Adkin, *The Bear Trap: Afghanistan's Untold Story*, London: Leo Cooper, 168.
5 Ibid, 111.
6 Ibid. 159.
7 General Shahnawaz Tani was from the neighboring town of Tani and enjoyed some popular support in the area. He later became DRA Defense Minister.
8 Omar and Haqani.
9 Perhaps this was a reconnaissance in force.
10 Local name for the chalk layers in the rock that mark this saddle.
11 Aleksandr Lyakhovskiy, *Tragediya i doblest' Afgana* [*The Tragedy and Valor of the Afghan Veterans*], Moscow: Iskona, 1995, 300-301.
12 General of the Army Valentin Leonidovich Varrenikov was head of the operations group of the Soviet Ministry of Defense in Afghanistan from 1985-1989.
13 Viktor Kutsenko, "Dzhavara" [Zhawar], *Soldat udachi* [*Soldier of Fortune*], July 1996, 24. *Soldat udachi* is affiliated with the American *Soldier of Fortune* magazine and does translate and print some articles on weapons and warfare from its parent magazine, but is an editorially-independent magazine that has enjoyed popularity in Russia and serves as a publishing outlet for veterans from Afghanistan, Chechnya, Ethiopia, Angola, and Mozambique. Some 60-percent of its articles are written by Russians.
14 Lyakhovskiy, 301.
15 Ibid.
16 S. Korennoy in Lyakhovskiy, 307-308.
17 Kutsenko, 24.
18 The 1st and 3d battalions of the 191 Separate Motorized Rifle Regiment stationed in Gazni.
19 Lyakhovksiy, 301.
20 S. Korennoy in Lyakhovski, 306-309.
21 The maximum size this force could have been was sixty men, since the Mi-8 could lift ten fully-equipped men at this altitude.
22 Kutsenko, 24-25.
23 Interviews with Lieutenant Omar, Mawlawi Nezamuddin Haqani, and Mawlawi Abdul-Rahman in Jalali and Grau, Chapter 11.
24 Following Zhawar Two, the 38th Commando Brigade became the base of the newly-formed 2nd Division. Casualty figures vary. The Soviet figures are 312 landed of which 25 survived. Korennoy in Lyakhovksi, 309.
25 Most probably, these were KAB-1500L-PR laser-guided bombs launched from the SU-25 ground attack aircraft. The KAB-1500L-PR carries a 1100 kilogram (2425 pound) warhead and can penetrate two meters of reinforced concrete buried in 20 meters of soil. It has a

one-two meter CEP (circular error probable). Yuri Zuenko and Sergey Korostelev, *Boevye samolety Rossii* [*Russian Combat Aircraft*], Moscow: ELAKOS, 1994, 170-171.

26 Interviews with Lieutenant Omar, Mawlawi Nezamuddin Haqani and Mawlawi Abdul-Rahman in Jalali and Grau, Chapter 11.

27 Ibid.

28 Korennoy in Lyakhovksi, 309.

29 Ibid

30 Interviews with Lieutenant Omar, Mawlawi Nezamuddin Haqani and Mawlawi Abdul-Rahman in Jalali and Grau, Chapter 11.

31 Ibid. Spetsnaz is a Russian term for Special Forces used in long-range reconnaissance and rugged fighting.

32 The 1st and 3d Battalions of the 345th Separate Parachute Regiment stationed at Bagram, the 4th Battalion of the 56th Air Assault Brigade and another battalion of this brigade stationed at Gardez and 2d Battalion of the 191st Separate Motorized Rifle Regiment stationed in Ghazni. The airborne and air assault battalions flew in Khost airfield while the Motorized Rifle Battalion drove into Tani.

33 Lyakhovski, 301.

34 Kutsenko, 25.

35 Lyakhovski, 304.

36 Interviews with Lieutenant Omar, Mawlawi Nezamuddin Haqani and Mawlawi Abdul-Rahman in Jalali and Grau, Chapter 11.

37 Korennoy in Lyakhovski, 310. At the end of 1988, Colonel Aleksandr Rutskoy, who was by then the deputy to the 40th Army Commander for aviation, was flying in this same region and was again shot down-this time with an air-to-air missile. He landed in Pakistan and was captured, and subsequently ransomed by the Soviet government. He was decorated with the title of "Hero of the Soviet Union." Later, Rutskoy entered politics and became the Vice President of the Russian Federation. However, in October 1993, he lead the opposition to Russian Federation President Yeltsin. He was arrested. In February 1994, he was freed from imprisonment by decree of the Duma of the Russian Federation.

38 Yousaf and Adkin, 171 and interviews conducted by Ali Jalali with Mujahideen in the fall of 1996. Mr. Jalali's sources chose to remain anonymous since they were sworn to secrecy at the time.

39 Lyakhovski, 305.

40 Kutsenko, 25.

41 Interviews with Lieutenant Omar, Mawlawi Nezamuddin Haqani, and Mawlawi Abdul-Rahman in Jalali and Grau, Chapter 11.

42 Korennoy in Lyakhovski, 311.

43 Nan is an unleavened oval-shaped flatbread, ranging in size from a small to a regular pizza.

44 Kutsenko, 25-26.

45 Ibid, 26.

46 Ibid.

47 Ibid.

48 Interviews with Lieutenant Omar, Mawlawi Nezamuddin Haqani and Mawlawi Abdul-Rahman in Jalali and Grau, Chapter 11.

49 Veterans of Zhawar have proposed to Haqani that the caves be restored and kept as a museum so that 200 years from now, people can visit them and reflect on their heritage. From 7-9 May 1990, 47 major Mujahideen field commanders from all over Afghanistan met at Zhawar in the first united of all major commanders of all factions. The purpose was to chart post-Soviet military strategy. Zhawar was chosen for the conference for its symbolic importance. *Afghan Information Center Monthly Bulletin*, Peshawar, May 1990.

50 Lyakhovski, 302-303.

51 Ibid, 304.

RECONNAISSANCE

8

Mountain Reconnaissance Russian Style

Lieutenant Colonel Lester W. Grau, US Army (ret.) based on the work of Colonel Michael Panov

This article first appeared in the March-April 2010 edition of *Infantry*

Sensors are wonderful; unmanned reconnaissance planes are great. But sensors cannot detect all activity, particularly in the mountains. Heat sensors can be defeated with a piece of carpet or a space blanket. Motion sensors can be defeated by freezing in place, since many sensor platforms are noisy and readily detected. Sensors should not be the sole basis of tactical intelligence. Human intelligence, derived from the local populace, is an integral part of tactical intelligence. But the best tactical intelligence still comes from boots on the ground. Sensors can identify areas that require a closer look, but the eyes of the skilled scout are still the best way to know what is going on in the folds and recesses of the mountains.

Movement in the mountains is difficult. It is not just the climbing and the effects of thinner air at altitude. The mountains can be a very harsh and unforgiving environment. Observation is not always enhanced by altitude and the distances are difficult to gauge by the novice. Communications and supply are difficult. The following Russian article discusses mountain conditions and movement considerations while scouting in them.

You're Not On the Plains Here: Combat Experience of Reconnaissance Units in Mountainous Terrain

Colonel Mikhail Panov in *Armeisky Sbornik*, May 2007 1

How reconnaissance groups move around in mountainous terrain is very different from how they move on level ground. In the mountains, they will encounter rapid rivers, cliffs, impassable ravines, ranges, mountain passes, and icy and snowy slopes. The scouts will face rockfalls and ice or snow avalanches. Severe climatic conditions (hurricane-force winds, thunderstorms, gales) in mountainous areas can negatively impact a reconnaissance group's operational capability since changes in weather fatigue them and the hot sun of the mountains make it hard to rest normally during the daytime.

However, in spite of the scouts' enormous difficulties in the mountains, it is precisely the mountain conditions that provide maximum stealth for setting up camp and moving to combat locations. This material will deal with how to move about correctly in mountain terrain and put its advantages to maximum use.

To operate successfully in the mountains, scouts need to be specially trained in conditions that most closely approximate the conditions of the locality in which they will have to operate behind enemy lines. From the standpoint of ease of movement, mountain terrain can tentatively be divided into: foothills (600-1,800 meters above sea level), and mountain (1,800-3,000 m) and high mountain areas (3,000 and above). Although tentative, this division is important when evaluating mountain terrain for reconnaissance actions.

Frequent and drastic temperature change in the mountains produces phenomena that are very dangerous for scouts, which is why they have to be able to use different external signs to identify these phenomena and take timely safety measures. Above all, the scouts must be able to determine in a timely fashion that inclement weather—thunderstorms, gales, snowstorms, etc.— is approaching.

Each scout should be familiar with the mountain climate and be able to take timely protective measures.

Solar radiation is much stronger in mountains than on plains and it increases with elevation. Ultraviolet rays greatly affect the human body. Sunburn is possible. The sun's rays harm the retina, causing acute pain and at times even temporary blindness. Sun glasses should be used to protect the eyes. The face should be protected with wide-brimmed headgear or gauze covering mask; breaks and rests should be taken in the shade.

Thin air is one of the numerous difficulties that are encountered in mountain conditions. Scouts who have not been properly trained or acclimated to the elevation experience oxygen deprivation, which causes "altitude sickness," accompanied by breathlessness, headache, nausea, vomiting, and so on. Thin air weakens the joints of the arms and legs, which could easily result in dislocation of the leg or arm even from a minor fall.

Thunderstorms create the risk of being struck by lightning, especially when scouts are on the crest of a slope, hilltop, or ledge. When a thunderstorm is approaching, you should seek cover in a cave or snow pit. Do not get under single overhanging rocks. Large metallic articles should be set apart and covered during a thunderstorm.

If the situation does not allow the scouts to wait out the storm because they have to "shake off" an enemy tail, they should continue along icy or snowy slopes, where there is less danger of being struck by a bolt of lightning, although a new danger arises, that of ice or snow avalanches.

Thunderstorms are usually accompanied by heavy rain or snowstorms, which makes it difficult and very dangerous to move in the mountains because the grassy and rocky slopes become slippery. Rain can trigger rock falls and snow/ice avalanches. You need to move carefully in these conditions and keep a close watch on all sides. It is known that atmospheric discharges more commonly strike tall, solitary trees. To seek cover under them from a thunderstorm is to subject oneself to danger.

Snowfall in the mountains makes it difficult to take one's bearings and visually monitor the terrain, which could mean falling completely unexpectedly into a deep snow-covered fracture. Snowfall raises the risk of snow slides. Snow can penetrate clothes and cause cold weather injury. It makes sense to seek cover and wait out heavy snowfall.

Special care, belay and self-belay should be practiced if it is necessary to continue to move during a snowstorm. Snow slides are most likely on moderately steep slopes because snow does not usually stick on very steep slopes. There could be slides from smooth rocky slopes as well as from smooth slopes covered by tall grass. With thaw, rain, or a warm wind, the mass of snow that has accumulated in the mountains begins to melt and breaks away, forming wet snow slides. Slide-prone regions can be identified by dug-out channels, broken trees and shrubs, and large piles of snow at the base of a slope. These areas should be circumvented. If it is not possible to circumvent, the snow should be tested for firmness. The recommended method is to move along one route, following in each other's footsteps, in single file, with more distance (five-six meters) between the men.

If a scout does find himself in a slide, he should do everything to remain on the surface of the moving snow and immediately close his nose and mouth so as not to be suffocated by snow powder. If, despite his efforts, a scout is engulfed by snow, he should assume a vertical position and work vigorously to make a space for air around his mouth and chest, and then try to burrow to the snow surface by expanding the space.

The chief causes of ice avalanches are drastic temperature changes in the mountains and the weight of masses of ice. To be safe, areas of possible avalanches should be traversed early in the morning when packed ice stays in place. These areas should be negotiated quickly and one at a time.

Rock falls are most dangerous after sunset and in the first few hours after sunrise. Rock fall-prone areas can be identified by rock piles at the foot of slopes, visible furrows from tumbling rocks, and scree and dust on ledges. Dangerous areas should be negotiated quickly, one by one, moving from cover to cover and keeping an eye on the slopes above.

After protracted rain and abundant snow thaw in the mountains, the upper soil layer becomes waterlogged. In some areas there is a build-up of semi-liquid masses of water, sand, pebbles, dirt, rock fragments, and so on. These masses of dirt and stones [mud-rock flow] sometimes slide down the slopes and along the valleys. A mudslide usually moves slowly, but on occasion it can fall without warning, sweeping away everything in its path. It is easy to identify mudslide-prone areas because of the accumulation of dirt, stones, and scree in the mountain valleys and at the foot of slopes.

There are other difficulties in addition to the ones mentioned above, first and foremost orientation difficulties. It is difficult to take one's bearings in the mountains, even with a map and compass. When choosing a direction on a map, it should be borne in mind that distances measured on a map are roughly 8-10 percent less than in actuality because a map shows a projection rather than the actual distance; neither does it take into account possible deviations from the planned route.[2]

It is difficult and dangerous to march at night in mountainous areas, especially where there are no roads or trails. Moving along an unexplored path at night could cause casualties.

Prior to a march, the group commander should reconnoiter the route to determine where there could be rock falls or snow or ice avalanches, and where there is cover; how to get over or bypass the hardest sections; where and how to ford mountain rivers [gorges]; and where to set up day rests or temporary cover in a thunderstorm.

In addition, the group commander needs to map landmarks that can easily be used at night and figure out the distances to them, as well as adjust route times and map reference points.

Prior to leaving the base area, the group commander should task his deputy or the most experienced scout with carefully monitoring the surroundings for at least two hours, paying particular attention to the direction of the planned march. The group usually sets up an observation post for this purpose on a commanding height to which the scout goes in secret and carries out surveillance with an optical device.

The unique characteristics of marching in the mountains require correct regulation of meals and water intake. The meals of scouts operating in the mountains should be

Рис. 1. Страховка спуска палкой, лопаткой, веревкой.

Figure 1 Descent with stick, spade or rope belay.

substantial. With the major physical stress, they should receive a hot meal once or twice per day. A strict water intake schedule will keep the scouts combat ready and prevent "altitude sickness."

While on the move, water intake should be limited to small amounts drunk from a canteen. A little bit of salt should be added to the drinking water because water in the mountains lacks salt. It is categorically forbidden to consume ice or snow instead of water [as they lower core body temperature].

A successful mountain trek depends largely on preliminary preparation and the group commander's level of experience. When preparing for a mountain trek you should thoroughly examine your footgear, wash your feet and carefully smooth your socks or foot wraps so that they do not chafe your feet; you should lighten your load as much as possible, taking only vital necessities. You should place something soft between the load and your back, and pad the straps of the backpack (rucksack), using grass or moss.

You should breathe evenly while on the move, inhaling only through the nose and exhaling fully. You should not speak when climbing up the mountain and under no circumstances should you smoke. Brief three-five minute stops are usually taken to restore normal breathing rhythm.

Move at an even pace, bending slightly and not straining. When going uphill, tilt the body slightly forward and step on the entire sole without bobbing. When going downhill, tilt the body back and step on the heel so as not to slide or fall (figure 1).

On steep slopes, your feet should be wrapped in rope or wire, or you should wear specially-adapted footgear to prevent sliding. You should loosen your belt slightly

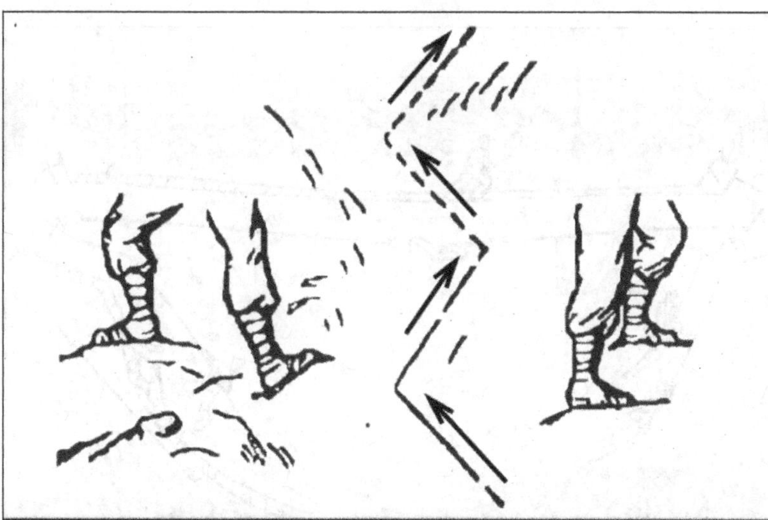

Рис. 2. *Постановка стоп « еАочкой » и « Аесенкой ».*

Figure 2 "Herringbone" and "Sidestep."

and undo your collar. Your step length should match the steepness of the slope. The steeper the climb, the shorter the step you should take. When going downhill, your step should be increased somewhat.

If the route is not along roads or tracks, you should walk uphill in a zigzag pattern rather than straight upward to make it easier, and "sidestep" or "herringbone" (figure 2). When going over unstable stones, talus, or narrow passes on a precipice, place your sole where there is a toehold and do not remove it until you have placed the other foot forward in a firm position.

Move carefully on steep rocky slopes so as not to kick downward rocks that could injury the scouts below. Use a stick on steep slopes. It is a good idea to carve out steps roughly 50 centimeters apart to make it easier to climb on steep, slippery, clayey, or icy slopes. Steps can be made by stamping in soft ground or in snow.

When going uphill on a grassy slope, your legs should be moved forward slightly, and the heavier the load and the steeper the slope, the further forward you should lean. When the climb is straight, your feet should be placed at an angle to each other in a "herringbone" pattern. The steeper the slope, the wider your foot angle should be, and the shorter the steps you take. Set your foot down on the entire sole. You should zigzag on long steep grassy slopes, and if the slope is sparsely covered with talus or rocks, you should move more tightly and not kick the talus downward.

All scouts should practice mountain techniques, including those whose units are deployed on level ground. They can use training gorges for this purpose (figure 3).

Рис. 3. *Учебный овраг для отработки приемов горной подготовки.*

Figure 3 Training gorge for practicing mountain techniques.

Colonel Panov's article emphasizes features of mountain movement and climate. Slow is fast in the mountains and steady, methodical movement, light packs, and acclimated physically-fit troops are necessary for mountain patrols. Effective patrols may last for days or weeks. Russian scouts are considered elite forces that perform missions that might be assigned to long-range reconnaissance patrols (LRRP) or special operations forces. Like US forces, scouts are assigned down to infantry battalion level.

Injured or wounded scouts are a problem in the mountains. Scouts frequently move at altitudes that are above helicopter ceilings. During the Soviet-Afghan War, the Soviets usually committed 12 troops to carry a single casualty down to the point a helicopter could reach the patient. Four men would carry, four would provide security and four would be ready to take over the carrying mission. A small number of casualties could quickly decimate a patrol.

Notes

1 Michael Panov, "Zdes' vam ne ravnina..." [You're not on the plains here..."], *Armeiskiy Sbornik* [*Army Digest*], May 2007, 24-26.

2 GPS does not always work particularly well in the mountains. Mountain configurations can cause over a 500-meter circular area probable (CEP). Do not trust GPS readings without confirming them on the map sheet.

ARTILLERY

9

Moving Artillery Forward: A Concept for the Fight in Afghanistan

Major Joseph A. Jackson, US Army

This article first appeared in the March 23, 2010 *Small Wars Journal* blog.

The United States Army is no stranger to mountainous and high-altitude war fighting. American history contains many instances of successfully executed mountain conflicts. Central to this success was the movement and use of artillery in direct support of those campaigns. The first notable American instance of moving artillery across mountainous terrain occurred when Colonel Henry Knox's Continental Army soldiers wheeled, sledged, and levered the guns from Fort Ticonderoga across the Berkshire Mountains in the winter of 1776. These fifty-nine assorted cannon became the deciding factor in General George Washington's siege of Boston. Other notable campaigns include the U.S. Army operations in the Italian Alps during World War II, the Taebaek Range of Korea, and the Annamite Range in Vietnam. Each of these locations and conditions provides ample instruction on artillery use in mountain warfare; yet this time fighting in the mountains of Afghanistan is proving to be a greater challenge than anticipated.

Strategists and commanders who consider employment of artillery in Afghanistan should take a fresh look at history, doctrine, and tactical concepts. Doing so will ensure artillery can employ optimally, and in sufficient strength, and of the correct caliber to create the tactical conditions for success. Without a significant increase in firepower delivered by a correspondingly lightweight and maneuverable field howitzer, the long-range fight in Afghanistan will devolve into an even deadlier and protracted conflict.

Solely relying on technology and precision munitions incrementally applied across the current arsenal will not achieve the conditions to exploit and pursue the insurgent fighters ever higher and farther into the mountains between Afghanistan and Pakistan. Operational planners, artillery professionals, congressional staffers, and military acquisition officers should examine these relevant histories, review doctrine, and consider their implications. These sources serve as a guide to develop successful and sustained operational approaches to combat the Afghan insurgency. They also provide a reference for adaptive tactics and procurement requirements for weapons needed in protracted high-altitude mountain warfare.

Defining the Operational Environment

In Afghanistan, the terrain and weather dictate the tactics and choice of weapons. Understanding the operational environment necessitates consideration of multiple factors. These dynamics include warfare in mountains that force non-linear fighting, training that does not prepare soldiers for vertical terrain, awkward and counterproductive positioning of the weapons, changing and treacherous weather conditions, and punishing temperatures that renders troops less effective.

In Afghanistan, significant mountain ranges such as the Himalayas and the Karakoram rise in the east. The Hindu Kush towers in the center of the country. The Suleman and Kirthar ranges jut toward the eastern border with Pakistan and extend into Baluchistan. Finally, the Paghman Range shrouds the capital city of Kabul. These ranges elevate more than two-thirds of Afghanistan's territory above 5,000 feet. These ranges provide natural concealment and protection for the insurgent fighters.

Fighting in extremely mountainous terrain and at high-altitudes is not linear. While forces move along pre-designated phase lines as on flat terrain, difficulties arise in maintaining continuity between units as they methodically scale from one point to the next. Fronts do not necessarily follow contiguous and sequential sets of ridges; they may even require simultaneous attacks on crests, ledges, and tactical objectives in opposite directions. A valley floor lying several thousand meters below may provide the only geographical point of continuity.[1]

Most armies train and equip themselves for conventional warfare on terrain that facilitates effective command and control and allows efficient employment of combined arms. Ideal terrain for mechanized forces are wide plains, rolling hills, plateaus, deserts, or sparsely populated regions that favor the linear and contiguous properties of maneuver warfare. None of these conditions is present in the bordering highlands along the eastern length of Afghanistan.

Extreme terrain also constrains fire support weapons. Artillery faces limitations imposed by steep road gradients and sharp bends that prevent deployment of the support vehicles and guns. To maintain the employment of guns as far forward as possible, batteries must disperse into sections, one or two guns per position, to maximize coverage to the supported units. To optimize the usefulness of the artillery, forces position their guns in terrain folds and on reverse slopes. Other positioning options include road heads, near villages, and along valleys. Deployment of artillery is often constrained because of logistics to support their use.[2]

Positioning of artillery becomes even more important when defending in mountainous terrain. The drastic changes in elevation and uneven ground make maintaining a continuous line of units tied together along their flanks difficult. A reverse-slope defense poses problems as well. While these positions mask unit movements and strengths, troops often lack sufficient overhead cover. Positions become susceptible to artillery fire and airbursts showering positions with fragmentation.[3]

Meteorological implications affect artillery use at higher elevations. Low air pressure, cold temperatures, and high wind speeds make standard firing tables ineffective.[4] These conditions increase inaccuracy. The lack of adequate maps and surveyed locations and the lack of precise meteorological reports increase the probability of error in range and altitude. Spotting rounds at high-altitude requires extra observers to walk rounds onto targets and to make drastic shifts to achieve accurate fire.

Human endurance must factor into the problem as well. Men cannot long endure temperatures ranging from as high as 128°F and as low as -15°F in the central highlands of Afghanistan and greater Southwest Asia. Prolonged exposure at high-altitudes depletes the strength of infantry units and requires frequent rotations of the troops.[5] This condition places greater responsibility for augmentation by the field artillery. Firepower must compensate for the aggregate reduction in troop strength. The tactical, geographical, and physical conditions interlock. Continued exploitation of the environmental conditions by the insurgents who are accustomed to these extremes allows them to engender more credibility than their weaponry and troop strength warrant.

Other weapons systems do not improve these circumstances. In fact, their limitations reinforce the demand for an artillery capability. Aircraft are of limited utility in high-altitude operations. Atmospheric conditions such as heavy rain, blizzards, fog, high winds, and low oxygen density limit performance. Camouflaged ground troops use the natural contours of the mountains that include deep shadows and overhanging ledges to prevent visual identification by aircraft. Aircraft use in valleys is dangerous; pre-positioned air defense weapons and massed small arms fire force aircraft to fly higher. This technique creates a visual positive identification problem and increases the risk of fratricide. Helicopters serve as good artillery spotters but weather and elevation limit their usefulness. Noise from approaching aircraft provides advance warning for units giving them time to hide among the rocks.[6]

The Current Paradigm

Presently, the US Army has implemented self-limiting measures in Afghanistan. This formidable institution refuses to commit its full spectrum of ground combat capabilities to overwhelm the enemy forces of the Taliban and Al Qaida. Instead, it continues to deploy its weapons in piece-meal fashion, arriving with a force that is too little too late. Nowhere is this more obvious than in the employment of the field artillery.

Delineating present limitations on the current artillery corps helps define the problem. Field Artillery battalions in support of expeditionary brigades continue to deploy with less than their full complement of cannons. Batteries often deploy with only 50-percent of their guns while the troops spend alternate periods serving as provisional infantry, quick reaction forces, augment logistics activities, and a myriad of training tasks focused on host nation capabilities. Additionally, the current arsenal lacks mobility. Cannons positioned on Forward Operating Bases arrive by helicopter. Once in position, they do not often reposition. U.S. artillery limits itself to only two calibers, 105-mm and 155-mm to engage targeted Taliban cells. Although these have proved effective in conventional wars in the mountains of Afghanistan, two is not enough. The other choice of weaponry, the M270, Multiple Launch Rocket System (MLRS), provides a significant capability and extreme precision. However, its optimal use fires at targets well beyond the immediate reach of the infantry involved in the fighting where individual and crew-served weapons make the difference.

Afghanistan presents a prolonged challenge. The restrictive practices of US forces coupled with creative Taliban tactics create an operational dilemma. Direct insurgent attacks against fortified positions pit enemy rockets, rocket-propelled grenades, and machine guns against mortars, and heavy and light crew-served weapons. The distance created by the insurgents using these weapons ensures that rifleman cannot effectively range them with their small arms. These tactics enable a small group of insurgents to attack and pin down a technologically advanced force in a fixed position. The result is that two divergent tactical fights emerge. The first fight belongs to the infantry, fighting in platoon or squad-sized actions at ranges of one kilometer or less. The second fight is in the purview of the artillery as it attempts to fight a much deeper battle against selected small targets out to 30 kilometers (see figure 1).

Taliban groups study and comprehend the US tactics. They observe that artillery remains fixed on bases and lacks mobility to follow the infantry into the deep defiles and higher elevations. After eight years of conflict, insurgent fighters further understand the limitations of shell fuse combinations and the restrictions that the environment places on rotary and fixed wing aircraft. By moving under cover of the

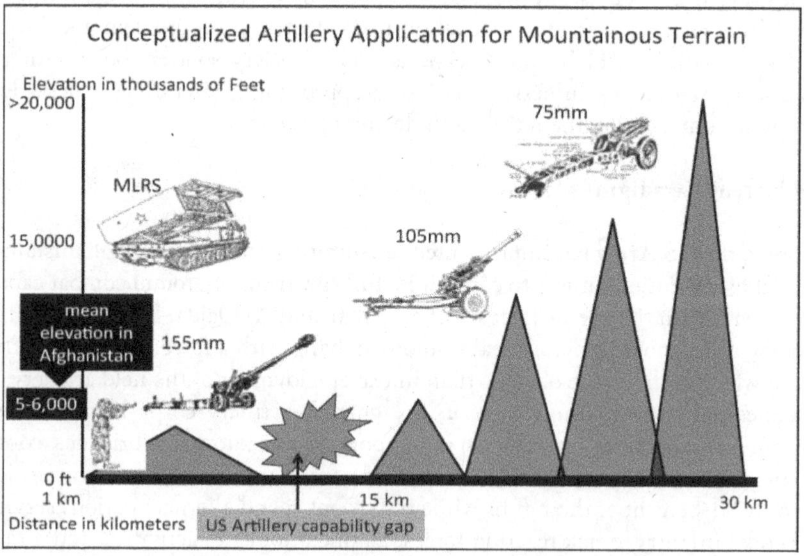

Figure 1 Field Artillery Application and Capability Gap in Afghanistan.[7]

mountains and along remote paths, they can avoid US radar and the limited number of forward observers who can engage them.

Insurgent fighters use the natural shape, strength, and remoteness of mountains to retain their freedom of maneuver and create the conditions for a long-range fight. They utilize the man-made caves that served them well during the Soviet conflict twenty years ago. Cave utilization forces heavily encumbered American soldiers to pursue the enemy into forbidding regions often beyond the range of direct support artillery. While mortars may provide a quick response, they still lack the punch, mass, and range to support the fight in Afghanistan for the long-term. This lack limits the ability of offensive forces to achieve two important aspects of offensive operations—exploitation and pursuit. The inadequate application of combat power over the past six years facilitated the increase in hostilities by the Taliban (see figure 2).

As recently as 2009, improved insurgent tactics now include firing volleys of rocket-propelled grenades, mortar rounds, and missiles from the back of trucks to allow insurgent groups to maneuver to disrupt coalition forces and seize key objectives such as remote outposts and towns. The lack of coalition troops in any given area and a corresponding lack of artillery to mitigate that deficiency in troop density has allowed the insurgency to fight along increasingly conventional lines not witnessed since Operation Anaconda in 2002.[8] The tactical dilemma that presented itself then emerges again now. The weapon system designed to engage in the long-range fight

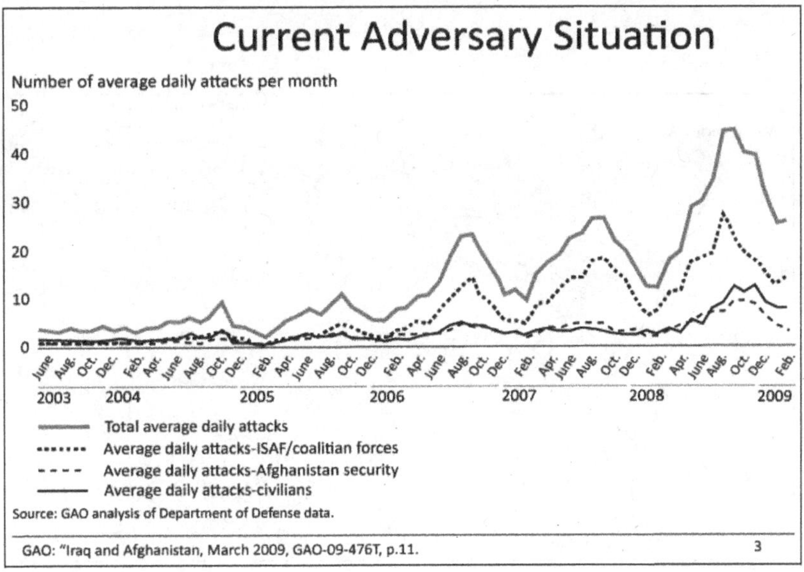

Figure 2 Location and Percentage of Taliban Attacks by Region in Afghanistan. Source: School of Advanced Military Studies, Exercise One. Data Extracted from U.S. Government Accounting Office Report on Iraq and Afghanistan, March 2009.

(cannons, howitzers, and rockets) remains noticeably absent from the majority of the fighting in significant numbers.

The Lens of History

The Soviet occupation and American approaches to fighting in Afghanistan warrant attention and denote significant but important contrasts. For the Soviet Union's five and two-thirds divisions, the geographical and operational limitations of Afghanistan reinforced the Soviet reliance on artillery as the centerpiece of their army formations. In contrast, Afghanistan has been an example of limited incremental technological application for the United States.

The complexity of fighting in Afghanistan produced an arguably counter-intuitive response for the Soviets. At first glance, the task of fighting an asymmetric enemy in largely uncharted territory would seem to warrant limited artillery formations in favor of lighter and more mobile forces. However, the experience of the Soviet artillery corps in the prosecution of the army's campaigns clearly noted that it remained a central combat arm in counterinsurgency warfare. Despite the limited maneuver space, winding mountain roads, and narrow valleys, creative methods of utilization allowed the artillery in certain circumstances to fight with limited or no infantry support. Moreover, the Soviet infantry fully appreciated the necessity for maintaining adequate fire support assets on all types of missions given the limited numbers of infantry battalions attempting to cover the entire country.

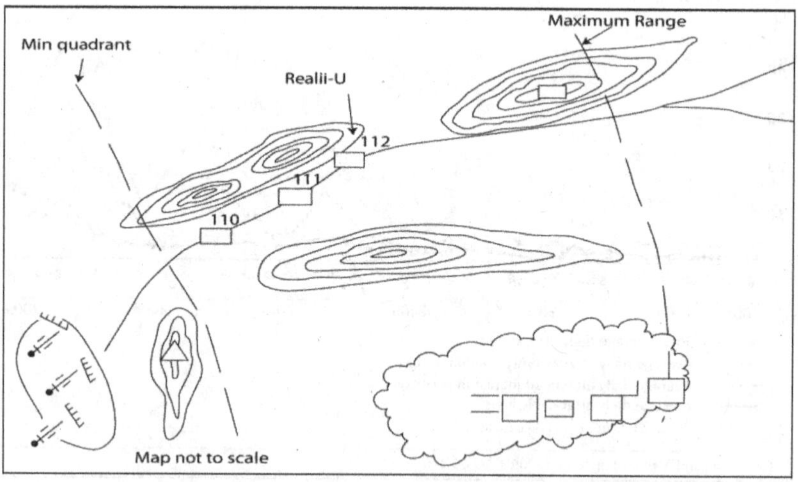

Figure 3 Diagram of the Artillery Ambush near Talukan, Afghanistan February 1986. Source: Foreign Military Studies Office, Fort Leavenworth, Kansas. http://www.ciaonet.org/cbr/cbr00/video/cbr_ctd/cbr_ctd_51.html (accessed 10 August 2009).

The Soviet Artillery Corps worked to integrate their weapons systems into the overall operational plan, rather than work to the exclusion of other arms. Mortars in significant numbers shared battle space with howitzers, aircraft, and cannons in large numbers synchronized their efforts against objectives, the Soviets sought to integrate as many weapons systems of as many calibers as possible to execute a combat mission. The Soviets did not permit the Mujahedeen's tactical exploitation of the environment to dictate the terms of the utility of rockets and howitzers, both towed and self-propelled systems proved useful. The decentralization of the artillery to support infantry platoon, company, and battalion level operations also revealed that Soviet Army officers became more adaptive and innovative over time. Though initially resistant to change, the Soviet Army proved increasingly flexible and adaptive out of tactical necessity.

Two notable examples illustrate the effectiveness of the Soviet artillery in Afghanistan. First, the successful artillery ambush conducted by Lieutenant V. Kozhbergenov, a D30 (122-mm) howitzer platoon leader, displays the accuracy that Soviet artillery operating in decentralized platoons could achieve through indirect and unobserved fire planning (figure 3). At night, the *Realii-U* operator, (a seismic recording device), reported that 10-15 people, two trucks and pack animals were passing through the eastern most target area, number 112. The platoon leader fired three concentrations. As the Mujahedeen continued to approach concentration 111, the gunners fired a volley. Then, the first piece switched to fire concentration 110 and the third piece switched to fire concentration 112. Number 2 gun continued to fire on concentration 111. The platoon destroyed two Toyota trucks, four pack animals and six men as well as destroying small arms and ammunition.[9]

In a second example, the Soviets made maximum use of a battalion of 152-mm 2S3 howitzers in direct fire mode in 1986 in the Baghlan Province of Afghanistan. A battalion of self-propelled howitzers employed direct fire across a wide front and moved forward by coordinated bounds by battery. Each successive bound brought the guns between 100 and 150 meters closer to their targets. The supporting towed 122-mm D30 artillery battery groups continued to fire concentrations across a one-and-a-half kilometer front and a depth of three kilometers. These concentrations effectively pinned the enemy inside the valley and the fortified villages, while the maneuvering 152-mm batteries systematically reduced specific enemy targets. By matching the sequencing of the fire plan to map lines, the guns accurately shifted fires from one line to the next destroying the enemy-covered firing positions.[10] The integrated fire plan successfully suppressed the opposition and enabled the capture of the village with limited exposure of Soviet troops to enemy fire. It eliminated the village stronghold.

For the Soviet Army changing tactics and fully integrating their indirect firing systems into the fighting gave them an increased level of success. The final withdrawal from Afghanistan rested not on faulty tactics. The Soviet political leadership in Moscow realized that the continued cost in lives and materiel could not substantiate or guarantee a clear political victory.[12]

The Artillery Gap of Operation Anaconda

In 2002, the United States committed troops to fight in Paktia province. Ironically, this had also been the location of brutal fighting between the Soviets and the mujahedeen during the Soviet-Afghan War. The Shahi-Kot Valley, the "Place of the King" had historically been a refuge for Afghan guerrillas. The valley became the scene for one of the largest battles of the US-Afghan War.

The fighting in the Shah-i-Kot valley rested on two inferences. First, that Al Qaida would not stand and fight, and second, that those weapons systems at hand (mortars and fixed and rotary-wing aircraft) would be sufficient to handle any fighting that did occur. In the months prior to Operation Anaconda, the US military, coalition special operations troops, and local Afghan militias banded together and defeated Al Qaida and elements of the Taliban in their attempt to control of the city of Kandahar and a few weeks later at Tora Bora.[12]

The United States Army did not recognize the fallacy in those assumptions until after fighting in the Shah-i-Kot Valley commenced. During Operation Anaconda, 1-18 March 2002, unlike the previous Soviet intervention, no artillery was present for the coalition and American troops. This absence of artillery created a noticeable capability gap that placed an increased burden on other weapons platforms such as mortars, helicopters, and an array of fixed-wing aircraft. Eventually, the application of mortars combined with air power destroyed large stores of enemy munitions, sealed off caves, spoiled a would-be counterattack, and scattered the survivors.[13]

Unfortunately, the notable achievements made by the use of coordinated close air support did not occur before significant delays, including fratricide, occurred in the original plan for the operation. Initially, the enemy retained the advantage in weapons and used them to disrupt the sweep through the valley. They not only possessed mortars of equal and greater caliber than the Americans, but also employed D30, 122-mm howitzers that could range the length of the valley floor.[14] In response, U.S. forces employed a number of aircraft including Apache AH-64 helicopters and fixed-wing aircraft that included A-10s, F-15Es, F-18s, and AC-130 gunships. These eventually bridged the gap between the limited mortar range and total lack of friendly artillery.

It is doubtful that the use of the current U.S. artillery arsenal in support of Operation Anaconda would have produced a decisive change in the immediate engagement. The overall inability of the artillery to support the mission due to a lack of expeditionary systems is the real point of significance. Of the two American systems that are air transportable (the 155-mm M198 and the 105-mm M119), only the M119 howitzer might have provided some measure of equal range and impact to offset the enemy D30s. The real value of Operation Anaconda to the artillery is that it illustrated how unprepared the US artillery arsenal was to fight an expeditionary war in Afghanistan's rugged landscape.

Future Possibilities

Fighting in Afghanistan reinforces the point that the fundamentals of artillery gunnery remain relevant regardless of the nature of the conflict and terrain. These recommendations serve as a basis of consideration and warning that while digital advancements in weaponry and precision munitions serve an important role, fighting in extremely mountainous terrain also requires the ability to move and to mass fire support at close ranges. During these engagements, aggregate rounds, not surgical precision establish a decisive advantage. Therefore, fundamental changes should match the demands of the operational environment.

Reforms must align people with devices, and that combination with the geography. To accomplish this, the US Army should consider expanding its arsenal of weapons systems and revising its doctrine. Specifically, the 75-mm pack howitzer is ideally suited for Afghanistan. This system, still in use by Pakistan and India along the Siachen Glacier and Kargil regions of the Kashmir, allows the artillery to move with the infantry into narrow defiles and up steep escarpments. It permits a high velocity, direct or indirect fire capability to destroy targets nested in caves and formidable terrain.

The 75-mm pack howitzer saw extensive wartime service with the US.Army issuing them to airborne and mountain units during World War II, (see figure 4). An airborne division, according to the organization of February 1944, had three 75-mm howitzer battalions. Glider units fielded two field artillery battalions that contained two 6-gun batteries each and one parachute field artillery battalion (three 4-gun batteries) totaling 36 pieces per division. In December 1944, new Tables of Organization and Equipment increased the divisional firepower to 60, 75-mm howitzers. The 10th Mountain Division contained three 75-mm howitzer battalions, containing 12 guns each.[15]

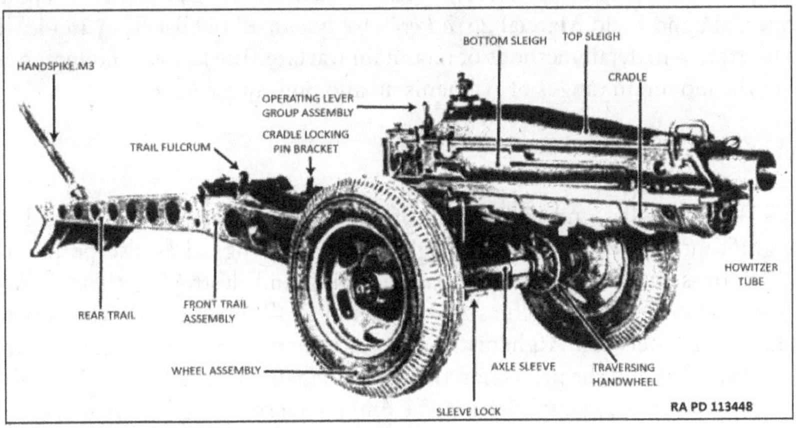

Figure 4: Diagram of the 75-mm Pack Howitzer from U.S. Army TM 9-319.[16]
Source: http://www.ww2gyrene.org/weapons_pack_howitzer.75mm.htm.

In the US Marine Corps, under the E-series Tables of Organization (TO), from 15 April 1943 divisional artillery included three 75-mm howitzer battalions of 12 guns each. The F-series TO from 5 May 1944 reduced the number of 75-mm battalions to two, and the G-series TO eventually removed them. The Marine Corps then shifted to 105-mm and 155-mm howitzers. Although the G-series TO was only adopted on 4 September 1945, in practice in some divisions the change was introduced early in 1945.[17]

The 75-mm howitzer, during its time of employment, proved successful. Nearly 5,000 guns were produced. As a part of Lend Lease, more than 800 saw service with the British forces in the Balkans where they proved excellent for fighting in that mountainous region. Today, Afghanistan provides another opportunity to use this weapon or something similar.[18]

As a practical matter, converting these largely ceremonial guns into operational howitzers should not be beyond consideration. The pack howitzer ranges out to 9 kilometers firing high explosive, anti-tank, and smoke rounds. It disassembles into six components and its total combat configured weight is 1,009 pounds. A concept for employment would not require replacement of the current 105-mm and 155-mm systems already present in Afghanistan. Instead, these weapons would augment the infantry with platoons or sections of pack howitzers as they pursue the Taliban into the highest points along the Afghan-Pakistan border.

Recommendations do not limit themselves to weapons systems. Training doctrine provides the link among the soldier, weapon, and the environment. Soldiers and leaders require access to core documents from which to draw relevant and expedient tactics and methods. A review of U.S. doctrinal history reveals that the army previously experienced fighting in mountainous terrain. The following points regarding texts and training deserve consideration. The Army should revise and publish Field Manual 70-10 *Mountain Operations*, Field Manual 70-15 *Operations in Snow and Extreme Cold*, and Field Manual 25-7 *Pack Operations*, all published in 1944.[19] These manuals explain in detail methods of mountain warfare. The lessons and methods still apply to the mountain ranges of Afghanistan and Southwest Asia.

Conclusion

Historical case studies and eight years of experience validate the need and utility of a significant artillery arsenal in Afghanistan. The rugged landscape provides a natural fortress for insurgents that make locating and destroying them difficult.[20] Transnational insurgencies such as Al Qaida and the Taliban exploit the network of mountain ranges between Afghanistan and Pakistan to their advantage.[21] Insurgents use the mountains and the protection they offer to maintain weapons parity with U.S. forces. American forces must adjust to the limits imparted by geographical and environmental conditions to offset the enemy advantage.

Mountain warfare dictates that combatants redefine their tactics and operational approaches to isolate and destroy these natural makeshift forts. Creating that isolation requires artillery. However, to fill that requirement the artillery must become mobile and directly support a limited number of infantry with a significant volume of fire to reduce and destroy these positions. The example of the Soviet Union's forces in Afghanistan points to the utility of using the full spectrum of indirect weapons systems from mortars to rockets, towed cannon, and self-propelled howitzers. They further reveal that field artillery can provide a useful and leading role in shaping operations and can directly defeat known insurgent defenses.

The United States Army's institutional memory remains short and neglects the fact that the field artillery proved effective in massing fires for decades in full-spectrum operations. Whether for offense, defense, or in deterring enemy forces, the artillery facilitated operational success in numerous contingency operations.[22] Maneuver commanders rely upon the presence of artillery to provide "firepower insurance"—having organic or assigned artillery capabilities present for any eventuality.

Certainly fighting in rugged terrain with artillery presents difficulties, but the case studies reveal that it is possible to use all types of artillery effectively and well beyond the confines of forward operating bases in Afghanistan and in the greater region of Southwest Asia. American forces need weapons that can destroy a smaller force in terrain not suitable for the current arsenal.

Major Joseph A. Jackson is a US Army Field Artillery officer with de-ployment and combat experiences in Bosnia, Korea, Iraq, and Afghanistan. He earned a bachelor's degree in history and Russian from Purdue University and master's degrees from the US Army Command and General Staff College and the School of Advanced Military Studies. Major Jackson is presently serving his second tour in Afghanistan with the NATO Training Mission, (NTM-A).

Notes

1 Lester Grau and Lieutenant Colonel Hernán Vázquez, Argentine Army. "Ground Combat at High Altitude," Foreign Military Studies Office, Fort Leavenworth, KS. http:// fmso.leavenworth.army.mil/products.htm#casia (accessed March 20, 2009) 4.

2 *Ibid.*, 9.

3 *Ibid.*, 5.

4 *Ibid.*, 6.

5 *Ibid.*, 6-9.

6 *Ibid.*, 6-9.

7 This diagram represents more than just the present mobility gap hindering combined infantry and artillery missions. It also represents how artillery might be echeloned by caliber according to altitude to fight in increasingly higher altitudes. The weapons systems displayed, with the exception of the 75mm pack howitzer, are currently in use by the U.S. Army. The Multiple Launch Rocket System typically engages targets beyond 20 kilometers and serves to fight the "deep fight" in American doctrine. Joseph A. Jackson, *Howitzers on High Ground: Considerations for Artillery in Southwest Asia*, School of Advanced Military Studies, (Fort Leavenworth, KS October 2009) 35.

8 Alfred de Montesquiou, "Marines launch new Afghan assault against Taliban," Associated Press, http://news.yahoo.com/s/ap/20090812/ap on re as/as afghanistan (accessed 12 August 2009).

9 Lester W. Grau, *Artillery and Counterinsurgency: The Soviet Experience in Afghanistan*, (Foreign Military Studies Office, Fort Leavenworth, Kansas, June 1997) 1-9.

10 Soviet artillery proved not only extremely valuable in the offensive examples above, but also in support of infantry breaking contact with the Mujahedeen. The supporting artillery fired suppressing fires in front of withdrawing troops to keep the enemy from closing or immediately reoccupying their fighting positions. Additional suppressive fires targeted key terrain along the flanks preventing the infantry from being enveloped or from receiving enfilading fires. The Russian General Staff, *The Soviet-Afghan War: How a Super Power Fought and Lost*. Translated and edited by Lester Grau and Michael A. Gress, (University of Kansas Press, 2002) 173-74.

11 Robert F. Baumann, *Russian-Soviet Unconventional Wars in the Caucasus, Central Asia, and Afghanistan*, (Washington: U.S. Army Center for Military History, 1993) 216.

12 The US Special Forces groups to included elements of forces from USSOCOM, JSOC and CIA's Special Activities Division. Other units and coalition partners consisted of TF 11, TF Bowie, and TF Dagger, British Royal Marines, and Canada's 3d Battalion, Princess Patricia's Canadian Light Infantry, and Joint Task Force 2. The German KSK, the Norwegian FSK and elements of the Australian Special Air Service Regiment, the New Zealand Special Air Service and Danish Special Forces from Jaegerkorpset and the Danish Frogman Corps and the Dutch Special Forces (Korps Commando Troepen) took part in Operation Anaconda. Steve Call, *Danger Close: Tactical Air Controllers in Afghanistan and Iraq*, (College Station: Texas A&M University Press, 2007) 59.

13 *Ibid.*, 78.

14 The decision not to deploy even light artillery—105-mm cannons—was deliberate. General Tommy Franks concluded that there was too little airlift capability to haul the pieces into the high altitudes. Instead, artillerymen of the 10th Mountain Division trained to operate the 120-mm mortar. This battle in the Shah-i-Kot Valley fueled the discussion on the use of mortars and airpower to fill the void where artillery pieces could provide a sufficient bridge in the tactical weapons system gap. The D30 122-mm howitzer possesses an effective range of 15.4 kilometers, with a rocket assisted projectile, this increases to 21.9 kilometers. In contrast, the 120mm mortar's maximum range is 7.6 kilometers. *OP 4: Field Artillery in Military Operations Other Than War: An Overview of the US Experience*, (Combat Studies Institute, CSI Press, Fort Leavenworth, Kansas) 37.

15 The U.S. Army recently decided to use the XVIII Airborne Corps and its aligned divisions consisting of the 82d Airborne Division, 101st Air Assault Division, and the 10th Mountain Division to rotate through Afghanistan. These light divisions could readily adapt to integrate a lightweight howitzer such as the 75-mm. A suitable stockpile of these weapons could conceivably be left in theater while individual brigades would employ them. This would allow a consistent and more mobile artillery presence in each brigade area of operations and not increase the logistical cost of shipping more of their own assigned artillery of 105-mm or 155-mm to Afghanistan.

16 The 75mm pack howitzer is a mobile, general purpose, field artillery piece. It is manually operated single loaded, and uses fixed and semi fixed ammunition. The firing mechanism is a continuous pull (self-cocking type, actuated by pulling a lanyard. The recoil mechanism is a hydro pneumatic type, having a floating piston and a pneumatic respirator. It provided direct or indirect fire and could elevate to fire high angle fire to deliver plunging fire on a target to a range of 9 kilometers. Data extracted from TM 9-319 (*75mm Pack Howitzer*).

17 In addition to service with mule units, the 75mm howitzer became the main gun for glider and parachute artillery battalions in World War II. Airborne units so equipped won battle

honors at Bastogne, Normandy, and Arnhem. The chief difference between pack howitzers for mule use and those for airborne operations were the wheels. The glider units had spoke wheels and the parachute units had pneumatic tires. See http://ibiblio. net/pub/academic/history/marshall/military/mil hist inst/a/artyC.asc.

18 Other systems are readily available on the world markets ready for use. The Italian-made OTOBREDA 105-mm howitzer, manufactured by GIAT industries – disassembles into 11 components (a section can assemble it in three minutes) and is transportable by pack animals. It weighs 1,310 kilograms, or 2,880 pounds. This is a favorable savings in weight even when compared to the 4,690 pounds for the M119. The OTOBREDA fires nine different charges with a maximum range of 18.1 kilometers. Thirty countries presently field more than 3,000 of these systems worldwide. Tests and actual employment of this system in demanding environments such as Sweden, India, and Malaysia proved successful. Within the immediate area of Afghanistan, India's Ordnance Factory Board currently produces the India Field Gun (IFG), a 105 mm howitzer specifically designed for mountain deployment. It weighs 2350 Kg (5,170 pounds) with a maximum range of 17.5 kilometers and deploys by either truck or helicopter. For heavier calibers, GIAT Industries produces the Caesar 155-mm self-propelled howitzer. This system integrates the prime mover and gun. The gun rests on the truck bed. The design specifically addresses the inability of towed 155-mm howitzers to deploy into mountainous areas. Sengupta, Prasun K., "105-mm Guns for Rapid Deployment Force and Mountain Warfare," *Asian Defence Journal*, no. 3 (March 1999) 24.

19 Some reprinted field manuals do exist. Nevertheless, they remain in the prevue of Special Operations forces. A significant portion of the army experienced fighting in rugged terrain well beyond the units comprising the traditional Special Forces community in Afghanistan since 2002. Previous army experience from World War II captured practical information of mountain troops in manuals such as, FM 25-7 *Pack Operations*, published in 1944. Presently published as FM 3-05 *Special Forces Use of Pack Animals*, Department of the Army, (Washington D.C., June 2004).

20 *Ibid.*, 1-2.

21 Bard E. O'Neill, *Insurgency and Terrorism: From Revolution to Apocalypse*, 2nd ed. (Washington D.C.: Potomac Books, Inc., 2005) 65-66.

22 *Ibid.*, 36-41.

Small Arms Fire

Lieutenant Colonel Lester W. Grau, US Army (ret.) based on the work of Colonel Michael Panov

Marksmanship ranges are located on flat land, firing positions are developed for flat land, firing tables are calculated for flat land. Things change when you shoot in the mountains. First off, it is much harder to determine range in the mountains, second, if you shoot uphill or downhill using flat-land marksmanship techniques, you will shoot over the target every time. Third, the wind, the temperature and the humidity is different in the mountains—and all play havoc with normal flat-land ballistics. In mountainous terrain like Afghanistan, the situation for the soldier or marine is compounded by the primary issue firearm—the M16 rifle or M4 carbine. These weapons fire a light-weight, high-velocity, short-range, flat-trajectory round in an area that requires a heavy-weight, medium-velocity, long-range, normal-trajectory round.

From the 1870s through the Korean War, US infantry trained to engage the enemy out past 900 yards and trained in the use of long-range plunging fire—a technique that works well in the mountains but does not work well with the 5.56-mm round. The current round's lethality falls off markedly at 200 meters (50 meters with the M4 carbine using standard M16 ball ammunition) and the M4 is sighted for maximum engagement at 300 meters.

Other countries have fought in the mountains and experienced many of the same problems. Russia has recently fought in the Caucasus Mountains and in the Hindu Kush and Suleiman Mountains of Afghanistan. They are working on many of the issues involved with shooting in the mountains. Russia's military comes from a different small arms tradition. They see the primary use of small arms fire as suppression of the enemy enabling the soldier to hold the enemy in place while artillery pummels him or allows the soldier to get close enough to dispatch him. The Kalashnikov assault rifle selector switch goes from "safe" to "automatic" to "semi-automatic." The US Marine Corps and Army come from a marksmanship tradition of hitting the enemy with single well-aimed shots, although they have gotten away from that tradition over the past three decades and more into suppressive fire. The M16 assault rifle selector switch goes from "safe" to "semi-automatic" to "automatic."

In May 2007, the Russian Army Digest *began their series on mountain warfare with the basics—firing positions.**

Note

* Colonel Mikhail Panov, "Strelyat'-tak pravil'no" ("The Right Way to Shoot"), in *Armeisky Sbornik* (*Army Digest*), May 2007, 19-21, and "Ochen' slozhno, no mozhno i nuzhno" ("Very difficult, but possible and necessary"), 22-23, in the same edition.

10

The Right Way to Shoot: How to Select Firing and Firing Ready Positions in Mountainous Conditions

Colonel Mikhail Panov in *Armeisky Sbornik*, May 2007

Mountainous terrain greatly impacts how well and effectively motorized rifle and tank units employ standard weapons (small arms). For example, there are a lot of unassailable (dead) spaces and hidden approaches that restrict the field of vision along the front as well as in depth. The varying rocky ground and vegetation also makes it much more difficult to locate and track targets. Indeed, wind direction and speed can change in a heartbeat depending on air temperature. Factor in that mountain climate is not at all uniform and extremely unstable, and that the air temperature varies considerably not only at different times of the year but even throughout the day, and sometimes even in the course of a few hours). It nowbecomes clear how important it is for soldiers to master the art of shooting accurately under such atypical conditions. In this issue of *Army Digest*, we offer just a few practical recommendations on training for firing and shooting in mountains. When preparing for battle in a mountainous area, remember that shooting accuracy is greatly influenced by the height difference between the firing position and the targets, shooting method, shooter position, limited ability to track targets on ridges, and other factors.

Personnel should acquire basic weapons skills during weapons training and range firing on level ground. Their shooting knowledge, proficiency and skills should then be improved at weapons training camps in the mountains.

Mountain marksmanship training should focus not only on firing position selection but also on the firing stance. The correct firing stance for small arms fire is critical because it stabilizes the weapon, thereby ensuring a hit.

Рис. 1. Положение для стрельбы из автомата лежа вниз со склона, обращенного к цели.

Figure 1 Prone position for firing an automatic rifle down a slope facing the target.

When considering a particular firing stance, it is important to remember that mountain conditions require that the position must be adjusted which, in turn, introduces new considerations or even different parts of the target into the ready position.

Let me remind you that a firing stance includes assuming a firing position and loading the weapon. Since there are no special techniques of weapon loading in the mountains, with the exception of the grenade launcher operator, let us take a closer look at position assumption. Let me point out straight away that the position will depend on the selected site (sloping or level ground) and the direction of fire (upward or downward). The firing position assumption technique is the same as on level ground, so we will consider firing stance (body, arm and leg position).

Shooting downward from a slope facing the target

Prone Position

The body is turned left and sideways at a 90° angle toward the plane of fire; the left leg is laid down against the slope in the direction of the target and the inner foot is thrust into a prepared hollow space (protrusion); the left elbow is thrust into the ground; the elbow angle is much larger than usual (120-160°); the right elbow is held close to the body. An automatic rifle is held by the front grip (Figure 1); a machine gun's bipod is thrust into the ground and the gun is held by the butt (figure 2).

Рис. 2. Положение для стрельбы из ручного пулемета лежа вниз со склона, обращенного к цели.

Figure 2 Prone position for firing a light machine gun down a slope facing the target.

When a soldier is firing a grenade launcher, his loader is on his right at of his outstretched left arm from the grenade launcher tube. The loader's body is turned right sideways toward the target; the right leg is laid down against the slope; the grenade pack is placed against the slope below the body. To load the launcher, the loader passes the grenade over the muzzle, holding it at the top with his left hand keeping the propellant motor at a palm's distance from the nozzle with the pin up.

The grenade launcher operator takes the grenade with his left hand from below by the nozzle and inserts it into the tube.

Рис. 3. Положение для стрельбы из автомата с колена вниз со склона,обращенного к цели.

Рис. 4. Положение для стрельбы из автомата сидя вниз с крутого склона, обращенного к цели.

Рис. 5. Положение для стрельбы из автомата сидя вниз с некрутого склона обращенного к цели.

Figure 3 Kneeling position for firing an automatic rifle down a slope facing the target.

Figure 4 Sitting position for firing an automatic rifle down a steep slope facing the target.

Figure 5 Standing position for firing an automatic rifle down a gentle slope facing the target.

This arrangement ensures complete coordination between the grenade launcher operator and his assistant, enables rapid loading, and allows the loader to adjust fire.

Kneeling Position

The right leg is turned in the direction of fire at an almost 90° angle; the foot is turned so that the right side lies on the ground; the left leg is laid down against the slope in the direction of fire, and the heel is thrust into a hollow space (protrusion). The shooter's right buttock is on the ground, and the left is thrust against the right heel (the heel keeps him from sliding down); the left elbow is thrust against the left kneecap; the right elbow is in the same position as for shooting on level ground; the automatic rifle (machine gun) is held by the front grip (figure 3).

Sitting Position

The shooter can fire from two positions, depending on the steepness of the slope. For example, if the slope is very steep (figure 4), the body is barely turned in the direction of fire. The shooter sits on the ground with his legs downward against the slope in the direction of the target, digging his heels into hollow spaces (protrusions) a shoulder width apart; the left elbow rests on the inner left knee or on the thigh, a palm's length above the knee; the right elbow lies at the top of the right thigh; the automatic rifle (machine gun) is held by the front grip.

The shooter assumes a slightly different sitting position on a gentler slope (figure 5). He sits on the ground with his legs crossed and drawn up under him (Eastern-style); the right heel is a support under the left thigh; the left heel is thrust into a hollow space (protrusion) of the slope; the left elbow is thrust against the thigh; the right elbow is at an 80-90° angle to the body; the left hand holds the automatic rifle (machine gun) by the front grip.

Standing Position

The body is turned left sideways in the direction of fire with the feet as close as possible (a wide stance causes too much height difference, making the position uncomfortable); both feet are turned toward the plane of fire at 90º; the automatic rifle is held by the foregrip (figure 6).

Рис. 6. Положение для стрельбы из автомата стоя вниз со склона, обращенного к цели.

Figure 6 Standing position for firing an automatic rifle down a slope facing the target.

Shooting upward from a slope facing the shooter:

Prone Position

The body is turned slightly toward the target; the shooter leans slightly on his left side; the left leg is stretched downward against the slope and the toecap is thrust into a hollow space (protrusion) in the ground; the right leg is firmly bent at the knee and pulled under the right side of the stomach; the toecap is thrust into a hollow space (protrusion); the arm position is the same as for firing on level ground; an automatic rifle is held by the front grip (a machine gun – by the bipod). If this does not provide the necessary angle of fire, the rifle is held by the magazine body (figure 7).

When firing a machine gun, recommend that you place a support under the front grip to raise the barrel.

Kneeling Position

The toecap is thrust into a hollow space (protrusion); the right knee rests on the ground, the thigh is in a vertical position; the left elbow rests on the upper kneecap; the right elbow is held at an almost straight angle; the automatic rifle is held by the lower magazine case (figure 8). If this arm position does not provide the necessary angle of fire, the arm is straightened and stretched upward; the automatic rifle (Kalashnikov light machine gun) is held by the front grip.

Рис. 7. Положение для стрельбы из автомата лежа вверх со склона, обращенного к стреляющему.

Figure 7 Prone position for firing an automatic rifle up a slope facing the shooter.

Рис. 8. Положение для стрельбы из автомата с колена вверх со склона, обращенного к стреляющему.

Figure 8 Kneeling position for firing an automatic rifle up a slope facing the shooter.

Рис. 9. Положение для стрельбы из автомата стоя вверх с некрутого склона, обращенного к стреляющему.

Figure 9 Standing position for firing an automatic rifle up a steep slope facing the shooter.

Standing Position

The following position should be assumed when firing on a very steep slopes: the body is turned in the direction of fire at an almost 90° angle; the left leg is firmly bent at the knee, with the foot turned inward and thrust into a hollow space (protrusion) on the ground; the right foot stands parallel to the left one, the leg is straight at the knee; the left elbow is thrust into the left knee at the upper kneecap; the right elbow hangs down freely (the angle between it and the body is 30-45°); the automatic rifle is held by the lower magazine body (figure 9).

Рис. 10. Положение для стрельбы из автомата стоя вверх с некрутого склона, обращенного к стреляющему.

Figure 10 Standing position for firing an automatic rifle up a gentle slope facing the shooter.

On gentler slopes, body position is the same as on level ground, the width between the feet is minimal, the feet are parallel, the left leg is half-bent, and the right leg is straight; the left elbow is pressed against the left side of the chest and the right elbow hangs down freely; the automatic rifle is held by the front grip or the lower magazine body (figure 10).

11

It is Difficult—But Possible and Necessary—to Master the Skills and Techniques of Shooting in Mountains

Colonel Mikhail Panov in *Armeisky Sbornik*, May 2007

We have reviewed some firing stance techniques for shooting in mountains. Let us pause briefly to better understand that all of them (techniques) are determined by the terrain features as well as the best firing conditions. In addition, mountain combat experience has confirmed that the abovementioned firing positions are optimal. But what is the best thing to do if one has to shoot upward or downward from a level area in the mountains? Here, too, there are time-tested techniques. Let us review them.

Downward shooting position

Prone Position

The body is turned at a 15-25° angle in the direction of fire, the legs are slightly spread apart, the right leg is half-bent; the left elbow is down over the edge of the slope; the right elbow is further off to the side than when shooting on level ground; the left hand holds the front grip of the automatic rifle. This shooting position is assumed by automatic riflemen, machine gunners, grenade launcher gunner and snipers (figure 1). Kalashnikov machine gun and Kalashnikov light machine gun bipods are placed on the slope.

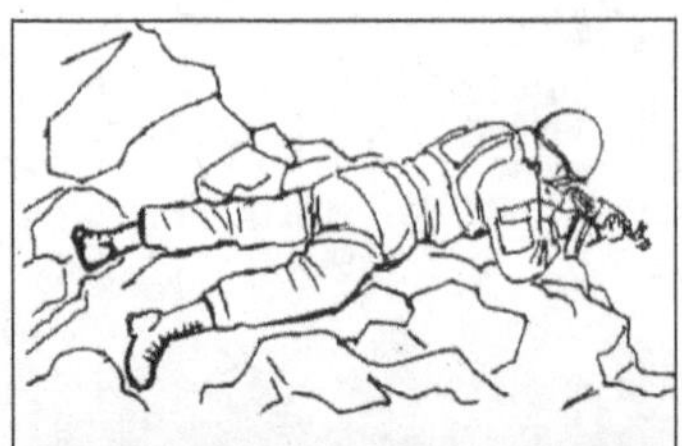

Рис. 1. Положение для стрельбы из автомата лежа с ровной площадки вниз.

Figure 1. Prone position for firing an automatic rifle downward from a level area.

Kneeling Position

The right leg is at a 60-70° angle to the plane of fire; the left leg is placed a half step forward, the lower leg is vertical, the foot is at a 45-60° angle to the plane of fire; the left elbow is lowered as close as possible to the left knee; the right elbow is held at a 75-90° angle to the body; the automatic rifle is held by the front grip (figure 2).

Рис. 2. Положение для стрельбы из автомата с колена ровной площадки вниз.

Figure 2 Kneeling position for firing downward from a level area.

Standing Position

The body is turned left sideways toward the plane of fire, the feet are no more than a shoulder width apart, the toecaps are turned slightly outwards; the left elbow rests on the hip; the right elbow is slightly lowered (45-60°) toward the body; the left hand holds the automatic rifle's front grip (figure 3).

Рис. 3. Положение для стрельбы из автомата стоя с ровной площадки вниз.

Figure 3 Standing position for firing downward from a level area.

Upward shooting position

Prone Position

Position remains the same as on level ground, except that the left hand holds the automatic rifle's magazine by the bottom (figure 4). If a support is used, it is higher than on level ground.

Рис. 4. Положение для стрельбы из автомата лежа с ровной площадки вверх.

Figure 4 Prone position for firing an automatic rifle upward from a level area.

Kneeling Position

The leg position is the same as for shooting downward but the right foot is completely turned, the right side of the heel touches the ground, the buttock rests on the inner foot rather than the back of the heel, the left elbow rests on the upper left knee; the left hand holds the automatic rifle's magazine by the bottom (figure 5a).

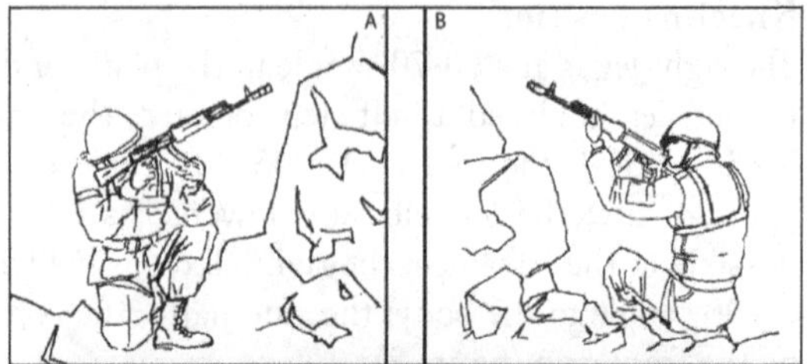

Рис. 5. Положение для стрельбы из автомата с колена с ровной площадки вверх: а — автомат удерживается левой рукой за нижнюю часть магазина;

Figure 5 Kneeling position for firing an automatic rifle upward from a level area: (A) The left hand holds the automatic rifle magazine by the bottom.

If this position does not provide the necessary angle of fire, the left elbow is not rested on the knee but is stretched upward, the elbow joint is straight, with the automatic rifle held by the front grip in this case (Figure 5b).

Standing Position

The standing position differs in only one respect: the left elbow rests on the left side of the chest rather than on the hip; the automatic rifle is held by the bottom of the magazine (figure 6a). If this position does not provide the necessary angle of fire, the left hand is removed from the support and stretched upward, the left elbow straight; the automatic rifle is held by the front grip (figure 6b).

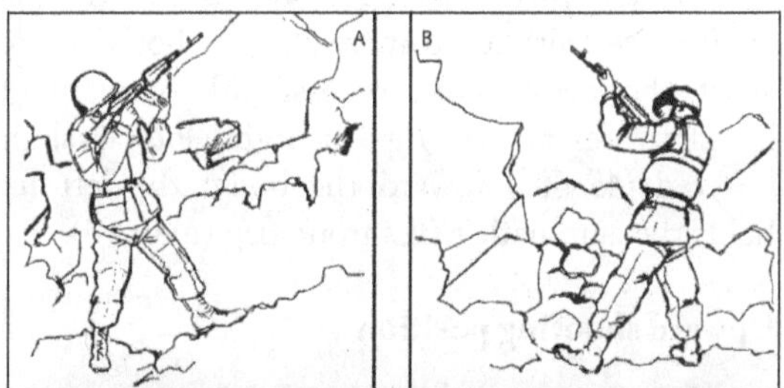

Рис. 6. Положение для стрельбы из автомата стоя с ровной площадки вверх: а — автомат удерживается левой рукой за нижнюю часть магазина;

Figure 6 Standing position for firing an automatic rifle upward from a level area: (A) The left hand holds the automatic rifle magazine by the bottom.

It should be pointed out that when shooting upward from a level area (road, bottom of a ravine or pit), a "kneeling" or "standing" position will be used most often since a "prone" position will be exposed and present the largest target.

Training in shooting along a slope should focus on the following: if the mountain peak is to the right, the left leg is lower down the slope and the inner boot side pushed against the slope, and the right leg is bent slightly and the outer boot rand thrust against the slope (figure 7a). If the mountain peak is to the left, the right leg is lower down the slope and the inner boot side is pushed against the slope, and the left leg is bent slightly and the outer boot side thrust against the slope (figure 7b).

All of these techniques ensure the most stable position and effective firing.

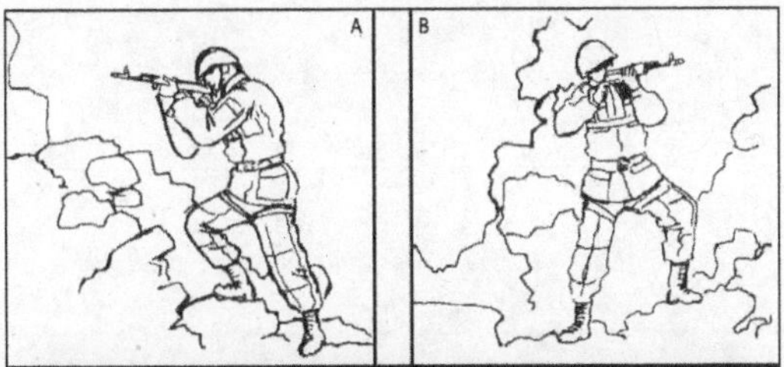

Рис. 7. Положение для стрельбы из автомата стоя вдоль склона:
а — при расположении склона справа; б — при расположении склона слева.

Figure 7 Standing position for firing an automatic rifle along a slope. (A) Slope to the right; (B) Slope to the left.

These techniques do not preclude other, more advantageous positions for shooting in the mountains. It all depends on the specific conditions: steepness of the slopes, soil conditions (icy, rocky, talus-strewn, grassy or snow-covered, creviced, uneven), weather (rain, snow).

Use of fractures on the terrain, stones, boulders and crevices as cover for shooters improves fire effectiveness and protects the shooter from enemy fire. Moreover, firing positions should be selected where there is less dead space. Large stones, shelf rocks, openings, rain channels, trees, shrubs, and so on should be used to cover the shooter.

In conclusion, firing position selection and ready positions need to be learned, and these techniques should be practiced at specially prepared sites that have slopes with varying steepness, large stones, shelf rocks, and so on.

These two articles may seem to be fairly basic and simple, and they are. But, for those who have not done mountain shooting, they contain good, time-saving information. Training for mountain firing should begin with a dry-fire stage where coaches and trainees move up and down mountains while identifying targets and taking up firing positions and stances to quickly and accurately engage the targets while using the terrain for protection from return fire. Once the trainee can automatically and smoothly assume the proper position and stance, he can begin live-fire training. Unlike conventional range fire, the coach and trainee move up and down and laterally in the mountains, identifying and engaging targets. Accuracy and speed of engagement are measured during this training and the tendency to overshoot the target is corrected. The trainee also conducts deliberate ricochet fire and plunging fire.

Once the trainees have mastered these measures, they should be reintegrated into squads for dry-fire maneuver training using bounding over-watch, flanking attacks and supporting and area fire while moving up, down and laterally in the mountains. Upon completion of this training, squads and platoons are ready to move into training against opposing forces (OPFOR) using MILES gear or simunitions.

SHELTER

12

So That Making Camp Makes You Stronger: Organizing Rest and Overnight Halts for Personnel in the Mountains

Colonel Michael Panov

Originally published in *Armeisky Sbornik*, May 2007*

It is important to get the soldiers into some sort of shelter for sleep and to escape storms. Naturally, the soldier's load prohibits carrying much more than a shelter half for that shelter, so the soldiers will need to use what nature provides.

The simplest shelter in a deeply snow-covered area is a snow hole. If there is no time to build comfortable shelter, it is necessary to dig a hole (cave), place brushwood, tree branches, grass, or moss in it, and close the entrance (with a shelter half, snow blocks). Here you can rest and wait out the foul weather. If the scouts expect to use this type of shelter for a long time, there should be space for beds, a fire, and gear, a hole for smoke, and ventilation.

You need to be careful when lighting a fire inside the shelter because beds and stocks of wood and brushwood could easily catch fire. You should also guard against fumes inside the shelter where the fire is burning.

Although it is warm inside the shelter, it is very dangerous to sleep on the bare ground or on snow, even in a sleeping bag. You must make a bed from small branches and fir twigs. To be warmer, the bed should be raised off the floor and at least 40-50 cm from the fire. If there are several people in the shelter, one of them should always keep watch to keep the fire going, guard the others while they rest, dry their shoes and clothes, and prepare a hot meal. He should also wake the other scouts from time to

* "Daby prival sil pribavlyal: Ob organizatsii otdykha lichnogo sostava v gorakh" ("So that making camp makes you stronger: Organizing rest and overnight halts for personnel in the mountains") in *Armeisky Sbornik* (*Army Digest*), May 2007, 27.

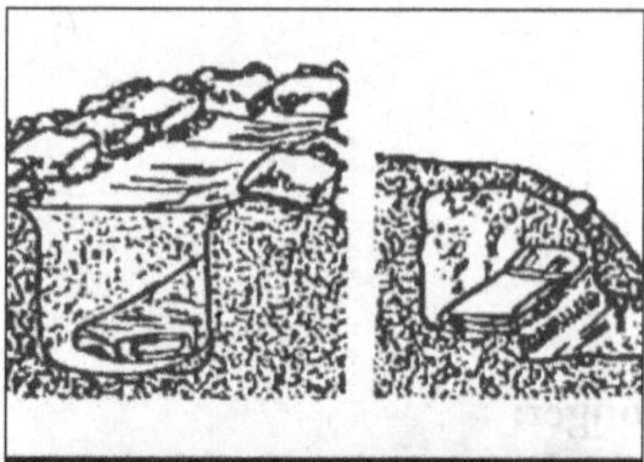

Рис. 1. Временные укрытия в глубоком снегу (слева - снежная яма, справа - пещера).

Figure 1 Temporary shelter in deep snow (left – a snow hole; right – a cave).

Рис. 2. Зимние временные укрытия типа чума (слева - из лыж и снега, справа - из жердей и плащ-палаток).

Figure 2 Rawhide-tent type temporary winter shelter (left – made of skis and snow; right – made of sticks and shelter halves).

time to change their sleep position because the part of the body facing the fire warms up while the other cools way down.

Snow holes and caves (figure 1) in high mountain areas can be used not only for temporary shelter for the scouts but also as the main base for a reconnaissance group if it is not possible to build an underground shelter or convert a cave into a base. Rawhide tents (figure 2) or canvas tents (figure 3) are more reliable temporary shelters where a fire can burn continually.

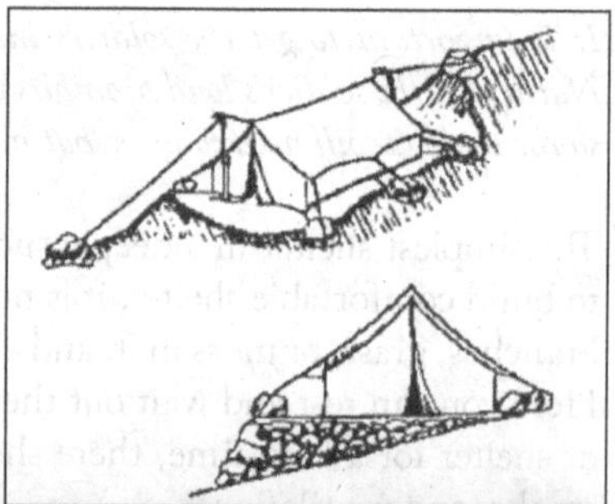

Рис. 3. Установка палатки в горной местности).

Figure 3 Tent setup in a mountain area.

It takes a lot of energy to build snow shelters, even though they are simple. To reduce perspiration, while working you should remove outerwear, loosen your belt, and remove gear. All snow and ice should be brushed off clothing and shoes before entering the shelter.

Probable rockfall areas should be avoided when selecting a base. It is also dangerous to set up a base in snow or ice avalanche-prone areas. Neither should you select dry riverbeds or areas where there could be mud flows (mudslides).

RUSSIAN ARMY TRAINING

Captain Charles K. Bartles, US Army

based on the work of Colonel Michael Panov

Russian/Soviet military traditions are very different than those encountered in Western/NATO armies. Despite efforts to modernize and develop a "contract" system of service similar to the system used in the United States, the Russian Armed Forces are still predominately conscript. The Russian draft process is a much dreaded event for most Russian men. In Russia, poor treatment and living conditions are synonymous with conscription. Conscripts are paid poorly by Russian standards, and their abuse is a well-known facet of military life in Russian society.

Unsurprisingly, conscription avoidance has become commonplace. Only about 10-percent of conscription age Russian males report for service in the Armed Forces every year even though there are many exemptions, deferments, and releases that will allow a potential conscript to legally avoid service. Often more illicit means are utilized to dodge the draft; bribery of the draft boards is common, as well simply not reporting for service. The end result of this process is a draft that is far from egalitarian. Wealthy and better educated Russian men (education deferments) rarely serve, while the vast majority of Russian conscripts come from poorer, less affluent families. This has caused great consternation in the military leadership due to missed recruiting quotas and the low quality of conscripts.

Training a conscript force presents many challenges to Russian commanders. Although conditions have improved in the Russian military in the last several years, the Russian military budget is still a fraction of Soviet levels. The priorities of the Russian defense budget are developing new weapon systems and paying salaries, not investing in the training of short term conscripts. Little time or money can be invested in soldiers that will only serve 12 months and the Russian Army has no centralized method of conscript training. Russian conscripts attend no "initial entry training" center, but are put into training platoons upon arrival at their units of assignment. In the Russian system, the unit commander (battalion level) has direct control over his conscripts' initial training, allowing training standards to vary widely between different units.

Russia's system of decentralized training presents a serious challenge for the commander of mountain troops. Any soldier operating in a mountain environment will quickly notice that the basic infantry skills (moving, shooting, and communicating) become quite difficult.

The Russian commander is faced with many difficulties training for mountain operations. Most of his conscripts are not the best and brightest, will cycle out every twelve months, and he has little in the way of specialized equipment to conduct training. Yet despite these difficulties Russia has enjoyed some success in mountain warfare. The Russian method presented below is Russia's approach to train short-term conscripts for the challenging task of mountain warfare.

The Russian Army approaches this challenge as it does most others, with the simplest solution possible. Graduates of the Marine Corps' Mountain Warfare Training Center, Bridgeport, California, and the US Army's Mountain Warfare School in Jericho, Vermont, may be surprised at how little emphasis is placed on land navigation and survival skills. In the Russian Army, Russian officers perform many duties performed by NCOs in Western/NATO armies and are trained to a much higher standard than their conscripts. Officer training is on par with many Western armies. Much of the advanced aspects of warfare that would normally be taught to all soldiers in Western/NATO armies are instead taught to only officers in the Russian Army. Land navigation, as well as many other duties, is an officers responsibility, little or no effort is made to train conscripts in this task. As for survival skills, one advantage of the Russian system of conscription is that recruits often serve in the same geographic area from which they are conscripted. Although the Russian conscript may not be familiar with the exact area in which he is serving, he may well already be accustomed to a mountain environment. The practice of assigning conscripts to areas around their homes further benefits the unit, as conscripts are demobilized and return to civilian life. In the event of national emergency, demobilized conscripts may return to duty with the same unit they demobilized from.

Although repetition is a common training strategy for most armies, the extent to which it is practiced in the Russian army will seem strange to those familiar with training in Western armies. Short service durations, no professional NCO corps, and little formalized training causes much less emphasis to be put on individual initiative at the tactical level. Instead Russian officers focus on training their troops on set drills and maneuvers. In tactical situations Russian officers pick from a "menu" of well rehearsed maneuvers. In the Russian view it far better to do several maneuvers very well, then trying ad-hoc tactics for situations as they develop.

Lessons for the US/NATO

The higher quality of training at US Army's mountain warfare schools is obvious, but there are still benefits to be had from studying Russian methods of training. These higher quality schools are available only to a small percentage of American troops. Most soldiers and marines deploying to Afghanistan do not receive mountain training and often these schools cater to the training of the infantry, with few slots for support personnel. The Global War on Terror has changed the military's emphasis from large scale conventional warfare, to low intensity conflict not characterized by defined fronts. There is no rear area and the "battle" can be anywhere in the Area of Operations. Consequently, Combat Support and Combat Service Support assets are in harm's way, as much as their Combat Arms brethren but, often without the same level of protection and training. The challenging conditions of mountain warfare and lack of availability of formalized training presents commanders with a difficult problem—how to conduct simple and effective training with a limited amount of resources. This problem is commonplace in the Russian army, and their solution to this problem may prove beneficial.

Perhaps the greatest benefit of these drills could be signatories to the NATO Partnership for Peace Program. This is a program for States of the former Soviet Union, which has focused on increasing security and democratic principles, while building a closer relationship with NATO.

Although many of these nations are adopting more US/NATO ideas into their armies, they are still patterned off their Soviet predecessor. There has been many changes in leadership and allegiances of the new armies of the former Soviet Union in the last 15 years, but the day-to-day lives of the officers have changed little. Generally, the conscription systems, role of NCOs, duties of officers, and resource constraints of the new armies of the former Soviet Union are still much closer to the Soviet/Russian model than to any of the traditional NATO members. The training methods discussed will be useful, familiar, and supportable to these armies.[*]

* Colonel Mikhail Panov, "Po sklonam i po tropam: Metodika obucheniya lichnogo sostava tekhnike preodoleniya gornykh prepyatsviy" ("Along Slopes and Tracks: Training Method for Overcoming Mountain Obstacles"), in *Armeisky Sbornik* (*Army Digest*), May 2007, 28-30.

13

Along Slopes and Tracks: Training Method for Overcoming Mountain Obstacles

Colonel Mikhail Panov in *Armeisky Sbornik*, May 2007

Negotiating Grassy Slopes

Objective: Teach how to negotiate grassy slopes with and without tracks and practice belay and self-belay with an ice axe.

Site: Rock-free grassy slopes of varying steepness overgrown by low grass, with and without tracks.

Outfit and Gear: Everyday uniform, boots, work gloves, climbing harness, ice axe, 20-tooth crampon per student.

Walking on tracks

Exercise 1. Single file, one step apart, climb 30-40 meters up a 20-30° track, turn around, and descend in the same formation to the initial position. Repeat two-three times.

Exercise 2. Same as Exercise 1 but with a full marching pack.

Ascending and descending grassy slopes

Exercise 1. Lined up abreast, facing the grassy slope (25-30°), two-three steps apart, "herringbone" up 15-20 meters, turn around via the right shoulder (so that the spike of the ice axe is toward the slope) and descend "straight forward" to the initial position. Repeat two-three times.

Exercise 2. Same as Exercise 1 but on a steeper slope.

Exercise 3. Single file, facing a 25-30° slope at a 45° angle and two-three steps apart, climb up 15-20 meters, turn around and descend to the initial position. Repeat two-four times.

Exercise 4. Single file climb 10-15 meters up a slope at a 45° angle, turn at a right angle to the path of ascent and "zigzag" up the same distance. Then turn around and descend by the same route to the initial position. Repeat two-three times.

Exercise 5. Single file, "zigzag" up a steeper slope with turns every five-seven meters, turn around and descend by the same route to the initial position. Repeat two-three times.

Crossing grassy slopes

Exercise 1. Single file, left shoulder to a 25-30° slope, two-three steps apart, go 20-30 meters across the slope, change the ice axe ready position, turn around via the left shoulder, and return to the initial position. Repeat two-four times.

Exercise 2. Same as Exercise 1 but on a steeper slope.

Self-arrest on grassy slopes

Exercise 1. Lined up and facing a 30-45° slope, two-three steps apart, "herringbone" up five-six meters, holding the ice axe on the slope side with the adze back and the spike forward. Lean forward, ice axe pick into the slope, then stick the adze into the slope, lie on the stomach and assume a comfortable position, holding the head of the ice axe with one hand and the shaft with the other, and brace spread legs against the slope (see figure). Stand up, turn around and descend to the initial position. Repeat two-four times.

Самозадержание на травянистом склоне при помощи ледоруба.

Exercise 2. While climbing up the slope (as in Exercise 1), execute a fall and self-arrest with an ice axe two-three times at a sudden command.

Exercise 3. Execute a fall and self-arrest with an ice axe several times at a sudden command while zigzagging up a slope in single file (at a 45° angle) two-three steps apart. Perform the exercise one-two times at a standstill, left (right) sideways to the slope, and then on the move.

Exercise 4. Same as Exercise 3 but while descending.

Exercise 5. Execute a fall and self-arrest with an ice axe several times at a sudden command while descending "straight forward" and lined up two-three steps apart. Repeat one-two times at a standstill, and then on the move.

Exercise 6. Single file, standing left (right) sideways to the slope, two-three steps apart, execute a fall and self-arrest at a standstill with an ice axe. Repeat two-four times.

Exercise 7. Execute a fall and self-arrest with an ice axe several times at a sudden command while crossing a slope left (right) sideways to the slope 20-30 meters apart. Repeat three-four times.

Methodical instruction

Once familiar with the technique, the students simultaneously execute the techniques and actions under the general command of the instructor.

When teaching grassy slope negotiation techniques, the instructor should make sure that the students place their feet correctly and position the ice axe with the pick toward the slope; the students should learn to use the ice axe right away for support and self-arrest when falling.

Some techniques should be learned without gear or weapons, first at in place and then on the move. The techniques should then be executed in full marching gear, moving at different rates on different slopes. While the techniques are being learned without gear or weapons, this equipment should be removed and placed where it cannot fall down and can be put on when needed.

Students should be taught self-arrest techniques with an ice axe after they have, for the most part, learned the rassy-slope negotiating technique.

The following is the training method for self-arrest with an ice axe at the instructor's sudden command: at the "Fall" command, the students self-arrest. It is important to learn how to move without kicking down loose stones. This way of walking teaches carefulness and precise movements.

In conclusion, the training is conducted in full marching gear and small arms. In addition to the "Fall" command, give the "Action" command while pointing in the direction of the enemy. At the same time, emphasize rapid and correct assumption of the ready position on a slope.

To prevent injury and accidents during grassy slope negotiating training there must be no:

- exercises on grassy slopes located under a rock mass where rockfall is possible or on a slope that ends in a steep cliff;
- simultaneous exercises by two groups located one above the other on the same grassy slope where there are loose individual stones;
- teaching of self-arrest techniques with an ice axe by falling on a grassy slope covered with rocks;

- ascent or descent on grassy slopes covered with loose stones without strict observance of the key rules of movement: keeping a certain distance, simultaneous turning by the entire group when zigzagging upwards, etc.

How to walk correctly on stones (talus)

Objective: Teach how to walk on small and large stones.

Site: Gentle slopes with small and large stones, old moraine, and a small rock pile.

Outfit and Gear: Everyday uniform, hiking boots, work gloves, ice axe.

Walking on small and large stones

Exercise 1. Line up one step apart, walk 15-20 meters on small stones and return to **the initial position. Repeat two-three times.**

Exercise 2. Walk 15-20 meters on large stones no more than a step apart and return to the initial position. Repeat two-three times at different speeds.

Exercise 3. Walk and run 15-20 meters on large stones slightly more than a step apart, landing on one foot in the initial position. Repeat two-three times at different speeds.

Exercise 4. Same as Exercises 1, 2 and 3 in full marching gear.

Methodical instruction

When teaching how to walk on stones, emphasize the placing the foot lightly, using the ice axe pick for support when off balance; and the danger of wedging a foot between large stones, and loose and fixed stones.

Negotiating talus-covered slopes

Objective: Teach the techniques of negotiating talus-covered slopes.

Site: Large, small, loose and fixed talus.

Outfit and Gear: Everyday uniform, hiking boots, gloves, ice axe.

Ascent and descent on talus

Exercise 1. Single file, two-three steps apart, climb 15-20 meters on large talus, turn around and descend to the initial position. Repeat one-two times.

Exercise 2. Single file, two-three steps apart, climb 20-25 meters on small packed talus, turn around and descend to the initial position. Repeat one-two times.

Exercise 3. Climb down 15-20 meters on small talus using the slipping of the talus itself – sliding on the feet, as it were - along with the sand and small stones, and gliding; return to the initial position taking small steps. Repeat two-three times at different speeds.

Exercise 4. Single file, two-three steps apart, climb 15-20 meters on small unstable talus, step onto hardened talus (small level surface area), turn around and descend to the initial position. Repeat two-three times.

Exercise 5. Climb 15-20 meters on unstable talus in full marching gear and run down to the initial position one by one. Repeat two-three times.

Moving across talus

Exercise 1. Single file, left sideways to the slope, two-three steps apart, go 15-20 meters across a slope with large talus. Turn around and go across the slope in the opposite direction, right sideways to the slope. Repeat two-three times.

Exercise 2. Same as Exercise 1 but on a slope with packed talus.

Exercise 3. Single file, cross a slope on unstable talus in full marching gear; after 15-20 meters, move to hardened talus or a grassy slope, turn around and cross the slope in the opposite direction. Repeat two-four times.

Methodical instruction

During ascent and descent on talus, make sure that the students are no more than two-three steps apart and that they immediately stop any stone movement.

To prevent injury and accidents when moving on talus, there must be no:

- exercises in rockfall-prone areas or on talus that ends in a steep cliff;
- simultaneous exercises of several groups in the same direction at different elevations;
- no group ascent or descent without strict observance of the basic rules of movement on talus;
- exercises held at sites that have not been inspected.

During exercises, it is necessary to watch the slopes and warn everyone there of a rockfall using the command "Rock."

14

Forward, Up, and then Down: How to Use a Rope to Climb a Steep Mountain Slope

Colonel Mikhail Panov in *Armeisky Sbornik*, May 2007*

Climbing using climbing stirrups

Objective: Teach how to climb on climbing stirrups.

Site: Two-three meter high cliffs and rock platforms.

Outfit and gear: Field uniforms, hiking boots, work gloves; two anchor ropes per squad; a backup rope (six meters) for each student.

Climbing on climbing stirrups with two anchor ropes

Exercise 1. Students gather near two or three sites that have been prepared for climbing with anchor ropes. One by one, they climb on climbing stirrups to the top of the two to three meter platform pulling and securing first one and then the other end of the anchor rope. Repeat two or three times.

Exercise 2. Same as Exercise 1 but in full field gear.

Methodical instruction

Carefully verify that the anchor rope is correctly secured for climbing with climbing stirrups.

To prevent injury and accidents, there must be no:

* Colonel Mikhail Panov, "Vpered, naverkh, a potom vniz: Kak c pomoshch'yu verevki podnyat'cya po krutomu sklonu gory" ("Forward, Up, and then Down: How to Use a Rope to Climb a Steep Mountain Slope"), in *Armeisky Sbornik* (*Army Digest*), May 2007, 31-36.

- start of training until it has been verified that the ropes are secured for climbing with climbing stirrups;
- ascent by the students by rope with stirrups until the inspector verifies that they have tied the knots correctly.

Rappelling

Objective: Teach sport, classic Dulfersitz and "figure eight" rappelling techniques.

Site: Grassy slopes that turn into cliffs, 45° rocky plates, eight to ten meters sheer rocky sections with outcroppings (terraces) above and below.

Outfit and Gear: Field uniform, hiking boots, work gloves; one 30 meters backup rope and two anchor ropes – per squad; a climbing harness, "figure eight" rappel device, two carabiners and one back-up rope (six meter) per student.

Sport rappelling down an anchor rope

Exercise 1. Students gather near anchor ropes. Then, in turn, holding the rope with both hands in a facing stance, they rappel adown a 35-40° grassy slope, pressing against the slope with legs positioned far apart. Gloved hands should slide (grip by turns) along the rope. Repeat the exercise two-three times.

Exercise 2. Same as Exercise 1 but in full field gear and self-belay with a Prusik knot.

Dulfersitz rappel with primary rope

Exercise 1. Secure the primary rope to a ledge or hook and throw the end down onto the grassy slope. Take turns passing the primary rope over the hips and body for a Dulfersitz rappel then removing it. Repeat three-five times.

Отработка суска по стене «дюльфером».

Exercise 2. Take turns rappelling down a 40-45° grassy slope on a Dulfersitz rope with belay from above.

Repeat two-three times.

Exercise 3. Take turns executing a Dulfersitz rappel down a 40-45° grassy slope with a "figure eight" and belay from above. Repeat two- three times.

Exercise 4. Same as Exercise 2 but with a Prusik knot for self-belaying.

Exercise S. Execute a Dulfersitz or "figure eight" rappel on an eight-ten meter rock wall in full field gear and with belay from above (figure). Repeat two-three times.

Exercise 6. Same as Exercise 5 but rappel down the rock wall in a pendulum traverse. Repeat two-three times.

Exercise 7. Execute a free rappel down a recessed rock wall; use a Prusik knot for self-belay in addition to a belay from above.

Exercise 8. Same as Exercise 7 but with two rappels in a row.

Methodological instruction

Students should be taught how to rappel on a primary rope in places where there are no rockfalls and where there are outcroppings (platforms) eight-ten meters above each other. The students take turns executing the exercise. Make sure that the hook has been securely hammered in, that the primary rope is properly anchored, etc.

Slope steepness should be gradually increased, ensuring belay from above. Before executing an exercise, the students should check the security of the hook and the primary rope. The students should be required to use a Prusik knot whenever there is the slightest danger of a fall during an exercise.

To prevent injury and accidents, there must be no:

- exercises on deteriorating cliffs or in rockfall-prone areas;
- rappel on a primary rope until the leader has checked that it is securely anchored to a ring placed a ledge or to a piton;
- rappel without belay from above or self-belay with a Prusik knot;
- no execution of an exercise by the next student until the inspector has checked that the rappel seat is secured and the knot has been correctly tied for belay and self-belay;
- unmonitored ascent to the platform for descent (see Addendum).

Conduct of training

Students begin this subject after they have learned the main rock climbing rules and self-belay and belay techniques, and have developed the necessary physical qualities and endurance.

The company commander organizes and conducts the lesson at the mountain training center's rock climbing training site. At the direction of the company commander, platoon commanders train the personnel in simple rock climbing techniques.

Before starting, the company commander checks that the personnel and mountain equipment are ready, and states the subject, training objectives, as well as the procedure for practicing the training points at the training sites. He then reminds the personnel of the general rock climbing safety requirements and takes the platoons around to the training sites and gives the command (signal) to start.

At training site No. 1, the platoon commander begins the lesson with practicing how to scale a rocky area of average difficulty while following the three anchor points rule. On a five to six meter rocky area, he selects three to four routes with overhangs, hollows, crevices and other unevenness for anchors and toeholds. He then explains and demonstrates how to execute it by personally completing one of the routes, after which he begins the practical training.

At the direction of the platoon commander (one per route), the students scale a rocky area and execute an easy rappel back to the initial position.

The platoon commander should follow the actions of all the students and correct their mistakes, focusing on correct and secure use of anchor points and toeholds, as well as well as watching the students' self-belay.

The platoon commander then moves on to training in ascent and sport rappel with a primary rope. He first checks the secured ropes and personally demonstrates the ascent and descent. He then decides the order (two students ascend, two stand in the initial position, two stand above, and the rest watch the ascent) and trains the platoon to ascend and then descend the two routes.

Once the students have mastered ascent and descent, the platoon commander moves on to intensive training in integrated ascent and descent using the flow-line method.

During this exercise, the platoon commander focuses on ensuring that the students correctly execute the footholds and rope pulls while ascending, and rappel in a strictly upright position, eliminating the possibility of "penduluming."

After practicing all exercise elements on the training site, the platoon commander does a brief review and moves to the next training site at the direction of the company commander.

At training site No. 2, the platoon commander begins by checking that the personnel have mastered the basic techniques of climbing a rocky area of average difficulty. When he is sure that they have done so, he moves on to how to scale a rocky ridge and crevice. If it turns out that some of the students have not mastered the climbing techniques (rules), the deputy platoon commander (mountain training instructor) provides them with additional training.

The platoon commander selects a slope with two or three horizontal or vertical ridges of varying form and steepness to teach how to negotiate a rocky ridge.

The platoon commander begins the rocky ridge negotiation training with a look at the obstacle and a demonstration of a technique for negotiating it, for example, sitting astride with shoulder belay from above. He then divides the students into shifts, decides the practice order (sequence) (two per ridge: one to execute the technique and the other to belay) and begins their practical training.

The students take their positions and prepare: the number ones for negotiating the ridge and the number twos for belaying, reporting to the platoon commander when they are ready. At the platoon commander's command, the ones negotiate a ridge three or four meters long sitting astride it, and the twos belay and watch the actions of the ones. After executing the exercise, the ones position themselves to belay and the belayers (the twos) rappel by an easy route and report their observations on the performance of the ones.

The platoon commander critiques the first shift and continues the training in the same order, checking the preparedness of each shift. When he is sure that the technique has been mastered, he moves on to practicing negotiating the ridge sideways in the same order.

Before practicing negotiating crevices, the platoon commander reminds the students that various forms and sizes of crevices make it much easier to negotiate difficult rocky routes because they have convenient supports and toeholds. He then reminds them of the safety requirements and explains and demonstrates how to negotiate a vertical crevice using a horizontal thrust with back belay from above and a piton. When he is sure that the students have correctly grasped the techniques, the platoon commander decides the exercise order (sequence) and moves on to practical training until all students are fully able to negotiate the crevice as well as belay.

In the same order, the students practice negotiating a horizontal crevice by crawling sideways with a piton belay.

After all techniques have been practiced, the platoon commander does a review and leads the platoon to training site No. 3.

Here, the company commander teaches how to negotiate difficult rocky areas (walls).

Two routes -- one five to six meters and the other 10 to 12 meters high -- are prepared for ascent and descent of a difficult rocky area (wall). The routes' relative positions and distance should make it possible to follow safety requirements while there is simultaneous training on both routes. The routes are checked and cleared of loose hanging rocks and rock debris ahead of time.

Following a brief explanation and tasking, the company commander checks the secured ropes and pitons, positions the platoon commander to belay, and demonstrates the technique by personally negotiating one of the routes. He then reminds the students of the safety requirements, advises them of the specifics of each route, decides the exercise order (sequence), and begins the practical training. Each student should first learn how to negotiate a five-six meter rock wall then perform a belay, after which he practices negotiating a 10-12 meter rock wall.

After teaching rock wall ascent, the company commander teaches in the same order how to perform a Dulfersitz rappel (with a "figure eight") and a pulley (carabiner) descent with belay from above.

During training to negotiate difficult rocky areas (walls), the company commander should focus on the correct use of various crags, cavities and cracks for toeholds and handholds, on the coordination of the belayer and the belayed, on following the belay and self-belay rules, as well as on the condition of the ropes, carabiners and pulleys.

The platoons change training sites 1, 2 and 3 every 50 minutes at the commander's command.

After all the scheduled training points have been practiced at sites 1, 2 and 3, the company commander at site No. 4 demonstrates how to set up a canyon crossing using a special cable, primary and backup ropes, pulleys and carabiners. He then demonstrates how to transport ammunition, crew-served weapons, food rations and gear, as well as people across a canyon.

After the practice, the company commander conducts an overall review and directs fixing particular weak points and preparing for the next lesson.

Addendum
SAMPLE

"approved"
Commander, Motorized Rifle Battalion
____
(rank, signature, last name)
________2009

LESSON PLAN
Scaling Rocky Slopes of Varying Steepness
________Motorized Rifle Company
________2009

SUBJECT:
"Scaling Rocky Areas of Varying Difficulty"

OBJECTIVES:
1. Train the company in the techniques of scaling rocky areas of varying difficulty and in self-belay.
2. Develop in the company personnel the physical and psychological qualities, endurance and stamina necessary for executing various tasks in the mountains.

TRAINING POINTS:
1 Scale rocky areas of average difficulty using various techniques.
2 Negotiate typical rocky terrain.
3 Scale difficult rocky areas (walls).
4 Move personnel, weapons and ammunition across a canyon.

TIME: 4 hours

PLACE: Alpinism training site, mountain training center.

TEXTBOOK AND GUIDE:
Training Units for Combat Operations in Mountain Regions (Moscow: Voyenizdat, 2007).

UNIFORM:
Summer mountain climbing gear and hiking boots, with weapons.

SUPPLIES:
Primary rope – two per squad, backup rope – one per squad, piton hammer – one per squad, pitons – 30, self-belay system, carabiner and work gloves – per student.

1. Preparatory Part – 10 minutes.
Check the training sites (mountain obstacles), personnel attendance, health (by asking), personnel and mountain gear preparedness; state the subject, training objectives and points, training procedure and time at the training sites; review general rock climbing safety requirements; move the platoons to the training sites and execute the preparatory exercises.

2. Main Part – 180 minutes.
Training Site No. 1
(50 minutes.)
(1st Motorized Rifle Platoon, Company Command)
1 Scale a rocky area of average difficulty using the three anchor points and self-belay rules.
2 Scale a rocky slope of average difficulty with a primary rope.
3 Execute a sport rappel. Instructor: Platoon Commander

Training Site No. 2
(50 minutes.)
(2nd Motorized Rifle Platoon, Flame Thrower and Anti-Tank Squad)
1 Scale a rocky ridge with belay from above.
2 Scale a rock chimney.

3 Scale a rocky area of average difficulty using three anchor points. Instructor: Platoon Commander

Training Site No. 3
(50 minutes.)
(3rd Motorized Rifle Platoon, Machine Gun and Grenade Launcher Squads)
1 Scale a difficult rocky area (wall) with belay from above.
2 Execute a Dulfersitz rappel (with a "figure eight").
3 Execute a pulley (carabiner) rappel. Instructor: Company Commander.

Training Site No. 4 (30 minutes.)
(Company Personnel)
1 Train to traverse a canyon using primary and backup ropes and pulleys (carabiners).
2 Move weapons and ammunition using a primary rope and pulleys (carabiners).

3 Move personnel on two primary ropes on a pulley (carabiner) using self-belay. Instructor: Company Commander

3. Concluding Part – 5 minutes.

Perform muscle relaxation exercises (while moving to the lesson review site); form the company, check that personnel, weapons and equipment are present; summarize the lesson: determine how well the techniques for scaling mountain obstacles have been learned, give grades, note the strengths and weaknesses; set tasks for eliminating the weaknesses and preparing for the next lesson.

Company Commander Work Procedure

1 Check the training site – on the day of the lesson before the company arrives.
2 Check that the company is prepared for the lessons, set tasks – five minutes.
3 Conduct the lesson at training site No. 3 – 150 minutes.
4 Check the organization and progress of the lessons at training sites Nos. 1 and 2 – 20 minutes. (at the start of each hour of study).
5 Conduct the lesson at training site No. 4 – 30 minutes.
6 Review and summarize the lesson – five minutes.

____ Motorized Rifle Company Commander

____

(rank, signature, last name)

15

Don't Let Me Down – Belay: Training Method for Negotiating Mountain Obstacles

Colonel Mikhail Panov in *Armeisky Sbornik*, May 2007*

Shoulder and back belay

Objective: Teach soldiers how to use the shoulder and back belay. **Site**: Grassy slopes of varying steepness with small outcropings and low cliffs.

Outfit and Gear: Field uniform, hiking boots and work gloves; primary rope—one for every two students; six meters backup rope, ice axe, and 20-tooth crampon per person.

Shoulder belay

Exercise 1. Students assume a facing stance in pairs on a 25-30° grassy slope. Each "pair" is tied together with a "leader's" knot. The first soldier of each "pair" descends six-eight meters, and the second assumes the correct shoulder belay position and feeds him the rope.

The first warns that he is going to check the belay and begins to jerk slightly and then more and more forcefully. The second feeds or tightens the rope to match the force of the jerking. The first then rappels the entire length of the rope and then climbs up. The second feeds or hauls up the rope in proportion to the descent and ascent. The students switch roles and repeat the exercise two or three times.

* Colonel Mikhail Panov, "Ne provedi, ctpakhovka: Metodika obycheniya voenno-slyzhashashchikh preodolehiyu prepyatstviy v gorakh" ("Don't Let Me Down, Belay: Training Method for Negotiating Mountain Obstacles"), in *Armeisky Sbornik* (*Army Digest*), May 2007, 37-40.

Exercise 2. Same as Exercise 1 but check the belay after giving warning, first with a bob, then with a jump, and finally with a sudden fall (figure 1).

Exercise 3. Paired together, the students begin to practice (by turns) assuming the appropriate belay position depending on the direction of the belayed's movement and the expected main jerk movement in the event of his fall. Repeat two or three times.

Exercise 4. Same as Exercise 3 but in full field gear.

Exercise 5. Train the students to stop a 60-kilogram log falling from a height of two meters.

Back belay

Exercise 1. Students gather in pairs on small platforms. Each "pair" is tied together with a rope. One serviceman seats himself comfortably on the platform and assumes the belay position, bracing his feet firmly. The other warns that he is going to check the belay and then does various jerks. They then switch roles. Repeat two or three times.

Exercise 2. Same as Exercise 1 but in full field gear. Jump instead of jerking to check the belay.

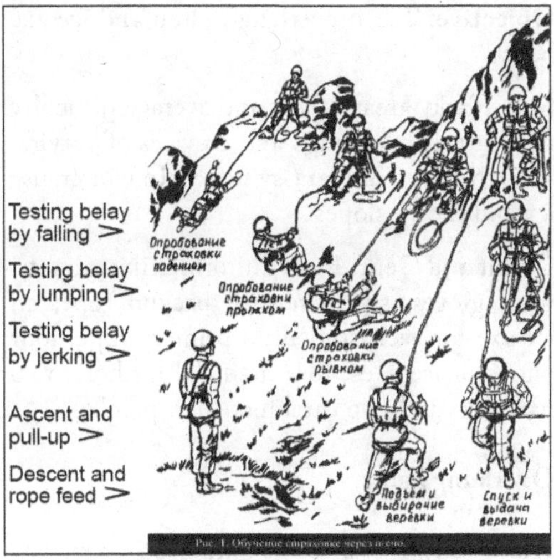

Figure 1 Shoulder belay training.

Methodological instruction

To familiarize the sol- diers with shoulder and back belay techniques, they are seated on a gentle slope. The instructor is located below them. He explains and demonstrates the techniques standing up (the best trained soldiers may also be used to demonstrate).

After a technique is demonstrated, the students execute it simultaneously on command. Some do so individually at the commander's instruction. The officer checks the execution and provides appropriate suggestions and explanations.

The commander must make sure that the students handle the rope properly, being especially watchful that the belayer assumes the correct position and has a good foot anchor, and that the belayed jerks and jumps to check the belay, gradually increasing the rope tension.

All students should be able to correctly determine the direction and force of the main jerk movement in the event that the belayed falls in order to quickly assume the necessary position and feed the rope to match the force of the jerk, thereby providing reliable belay.

To prevent injury and accidents, there must be no:

- lessons at rockfall-prone sites or on slopes close to precipices;
- belay training with jerks and falls on stone-covered grassy slopes;
- forceful jerks without making sure that the belayer is ready;
- belaying without work gloves.

Overhang piton and ice axe belay

Рис. 2. Обучение страховке через скальный крюк.

Figure 2 Instruction in piton belay.

Objective: Teach overhang, piton and ice axe belay.

Site: Rocky areas of low and average difficulty with sturdy overhangs and crevices of varying sizes that turn into grassy slopes. In winter, use icy and snowy slopes.

Outfit and Gear: Field uniform, hiking boots, work gloves; six meters of backup rope; an ice axe for each student; a primary rope, four pitons or ice axes, four artificial anchors, two hammers and two carabiners per "pair."

Overhang belay

Exercise 1. The students gather on overhangs in pairs. The second student in each "pair" throws the rope onto the overhang and positions himself for belay from above. The first student descends five-six meters, and the second feeds him rope. Then the first tests the belay with a jerk and hangs freely on the rope, holding his arms out to the sides. Then the first climbs up and the second pulls up the rope. The students then switch roles and repeat the same thing. Repeat two or three times.

Exercise 2. Same as Exercise 1 but the first student moves crosswise the slope instead of down, and the second positions himself to belay from the side.

Exercise 3. The first student moves up the slope at a 45° angle, and the second belays from below. Then the first rappels to the second by the same route to replace him. Repeat two or three times.

Exercise 4. Same as Exercises 1, 2 and 3 but in full field gear.

Piton belay

Exercise 1. The students gather by a rock wall in pairs (figure 2). The second student of each "pair" finds a crack, drives in a piton and checks that it is secure. He then clips a carabiner into the piton, inserts a rope in it and positions himself for belay from above after letting the first student know that he is ready. The first descends three to five meters (the second feeds him rope), tests the belay with a jerk and hangs freely on the rope, moving his arms out to the sides. The first then ascends to the second, who pulls up the rope. They switch roles and repeat the exercise. Repeat two-three times.

Exercise 2. Perform the exercise in full field gear.

Artificial Anchor Belay

Exercise 1. Students gather by a rock wall in pairs. The second student of each "pair" finds a crack, lays an appropriate anchor in it and checks that it is secure. He then clips a caribiner to the loop, inserts a rope in it, positions himself for belay from above, and lets the first student know that he is ready. The first student descends three to five meters (the second feeds him rope), tests the belay with a jerk and hangs freely on the rope, moving his arms out to the sides. The first then ascends to the second, who pulls up the rope. They switch roles and repeat the exercise. Repeat two or three times.

Exercise 2. Perform the exercise in full field gear.

Ice axe belay on an icy slope

Exercise 1. Students gather on an icy slope in pairs. The second student of each "pair" screws in an ice axe above his head, checks that it is secure, and covers it with snow. He then inserts a piton into a carabiner with the gate upward, inserts a rope in it, positions himself for belay from above, and lets the first student know that he is ready. The first student descends three to five meters (the second feeds him rope), tests the belay with a jerk and hangs freely on the rope, moving his arms out to the sides. The first then ascends to the second, who pulls up the rope. They switch roles and repeat the exercise. Repeat two or three times.

Exercise 2. Perform the exercise in full field gear.

Ice axe belay on a snowy slope

Exercise 1. Students line up in pairs on a 30-35° snowy slope five to six meters apart. The second student of each "pair" drives in an ice axe halfway, throws a rope on it, and positions himself for belay from above. The first student then descends five-six meters (the second feeds him rope), tests the belay with a jerk and ascends to his partner, who pulls up the rope. They switch roles and repeat the exercise. Repeat two to three times.

Exercise 2. Repeat the same exercise in full field gear.

Methodological instruction

Sites for overhang, piton and ice axe belay training should be carefully inspected and cleared of unstable rocks.

The instructor uses the best trained soldiers to demonstrate the belay techniques to the students.

After familiarizing themselves with the exercises, the students are paired up, receive their assignments, and move to their designated sites, where they perform the exercises.

The commander observes the exercises and critiques as necessary.

The overhangs used for belay should be carefully prepared. The students should test the piton, artificial anchor and ice axe belay to make sure that they are secure.

To prevent injury and accidents, there should be no:

- lessons at rockfall-prone sites or on slopes close to precipices;
- training with a primary rope that has not been pre-tested;
- independent exercises until the students are able to conduct a reliable belay;
- working in "pairs" without the instructor testing the belay (overhang prepared, piton properly secured, rope properly placed in the carabiner, etc.).

Prusik and Carabiner Belay (mass) with rope hand rail

Objective: Teach rope hand rail installation plus Prusik and carabiner belay.

Site: Grassy 30-40° slope and rocky areas of low to average difficulty with narrow terraces at heights of two meters and above.

Outfit and Gear: Field uniform; hiking boots and work gloves, backup rope, six meters climbing harness, carabiner and ice axe for each student; two primary ropes, four hammers, four pitons and four anchors per squad.

Exercise 1. Students line up in single files on anchored ropes with three to five meters between them. The students take turns approaching the primary rope, tie themselves to it with a Prusik knot, and climb three-five meters up the grassy slope, moving the knot along the rope. After testing the self-belay in the "hands down" position by stepping back and jumping, the students loosen the knot and continue to climb. In this way they all climb up one-by-one. The next student begins to tie himself to the rope when the previous student reaches the uppermost point of the secured rail and positions himself for self-belay or moves to a safe place. After the ascent, the students take turn rappelling back down using ice axe self-belay. Repeat two or three times.

Exercise 2. Students gather in a single file on a grassy slope sideways to a rock wall on which a rope has been anchored with carabiners and pitons driven in five or six meters apart. The students take turns approaching the primary rope, tie themselves to it with either a Prusik knot or a sliding carabiner, and traverse the slope, moving the knot or the carabiner along the slope. In the middle of the rope (between two pitons), they lower themselves and hang on the backup rope. They then ascend and continue to the end of the rope, after which they secure themselves to another rope with a carabiner and return to the starting point by another route. The next student in the squad begins to tie himself with a Prusik knot or sliding carabiner when the previous student has tested the self-belay and moved on. Repeat two-three times.

Exercise 3. Same as Exercise 3 but in full field gear. The primary rope is stretched one meter above a terrace path that is at least three meters above a grassy slope. Move lengthwise along the terrace rope using a Prusik knot or sliding carabiner self-belay (figure 3).

Figure 3 Training to use a Prusik knot on a rocky route.

Methodological instruction

The site selected to teach rail setup and Prusik knot or sliding carabiner self-belay, especially on rocky areas, should be carefully checked and cleared of unstable rocks.

The instructor should monitor the securing of the primary rope and the hand rail setup.

During the exercises, the instructor should ensure that the sequence and movement along the hand rails are strictly followed, and that the students first clip themselves to the next rail strand and then, and only then, unclip themselves from the previous one.

When one group of students has performed all the exercises, the next group trains at the same site using the rails that have been set up. It should be pointed out to the students that at difficult points along the rope they must remember to move the knot an outstretched arm's length away before moving on (the knot should always be smooth). All students should make sure that the Prusik self-belay is secure.

To avoid injury and accidents, there should be no:

- lessons at rockfall-prone sites or on slopes with loose rocks;
- exercises until the instructor tests the rails and the primary rope;
- movement along the rope until the student checks that the knot has been correctly tied.

16

How to "Tame" a River: Getting Across Water Obstacles in the Mountains

Colonel Mikhail Panov in *Armeisky Sbornik*, May 2007*

Fording

Objective: Teach how to ford mountain streams and rivers.

Site: Dry mountain riverbeds with rocks; rivers 50-60 centimeters deep, four-eight meters wide, with current velocity of up to three meters-per-second (m/sec).

Outfit and Gear: Field uniform, hiking boots, work gloves; three poles, three primary ropes per squad; climbing harness, backup rope (twice the length of the primary rope), one carabiner per student.

Exercise 1. Ford a knee-deep mountain river singly with pole self-belay and primary rope belay from the bank. Repeat one or two times.

Exercise 2. Ford a knee-deep mountain river in three ranks with hands on each other's shoulders. Repeat two-three times.

Exercise 3. Ford a mountain river in a circle of eight to ten people using hands on each other's shoulders. As it moves forward, the circle turns against the current. Repeat two to four times.

Exercise 4. Ford in a single file (eight to ten people) holding onto each other's belts. Repeat two-four times.

Exercise 5. Same as Exercises 2, 3, and 4 but in full field gear.

* Colonel Mikhail Panov, "Kak "ykrotit'" reku: Ymet' perepravlyat'cya cherez vodnye pregrady v gorakh" ("How to 'Tame' a River: Getting Across Water Obstacles in the Mountains"), in *Armeisky Sbornik* (*Army Digest*), May 2007, 41-45.

Exercise 6. Ford a 50-60 centimeter-deep mountain river singly with self-belay using a carabiner or a Prusik knot on a primary rope that is stretched diagonally from one bank to the other 30-40 centimeters above the water level. Students cross on the downstream side of the rope at a stretched arm's length from the primary rope. Repeat one-two times.

Training in the various fording methods should begin with exercises in an old mountain river bed (with no water) or on a four to eight meter area with stones.

Select a suitable river, determine its depth and current velocity, and physically check (using a belay) that it is fordable.

Set up rescue stations a short distance downstream from the fording spot to monitor the fording. Rescue "pairs" should be linked by rope so that one of them is ready to render aid while the other, positioned for self-belay, is ready to provide belay.

The instructor must be on the bank between those fording the river and the rescue station and personally direct the fording.

Ford in footgear and clothing. Make sure that the students do not lift their legs high when fording. Fording training should take place in sunny (warm) weather at the end of the training day.

To avoid injury and accidents during fording, there should be no:

- training at river sites that the commander has not personally inspected (where it has not been established ahead of time that the river is fordable);
- fording until the appropriate preparatory exercises have been practiced on land;
- fording without rescue posts;
- training in low water and air temperature or fording without footgear and uniform;
- fording with a carabiner or Prusik knot on a long leash from which it is not possible to reach the primary rope.

Water Crossing

Objective: Teach crossing of mountain rivers.

Site: Mountain rivers and rivers with scattered stones and high bluff banks.

Outfit and Gear: Everyday uniform, hiking boots, work gloves; three poles, three primary ropes per squad; climbing harness, backup rope twice as long as the primary rope, one carabiner per student.

Exercise 1. Negotiate a three-meter wide mountain river by jumping on stones 80-100 centimeters apart. Repeat two-three times.

Exercise 2. Same as Exercise 1 but with the stones further apart and with self-belay with a primary rope, stretched from bank to bank, to which a backup rope with carabiner is attached.

Exercise 3. Negotiate a mountain river by log-crossing with self-belay as in Exercise 2. Repeat two-three times.

Exercise 4. Take a load across a mountain river along two primary ropes using a pulley or carabiner. Repeat one-two times.

Exercise 5. Cross along two primary ropes using a climbing harness and pulley or carabiner. Repeat one-two times.

Exercise 6. Same as Exercises 1 and 3 but in full field gear.

Methodical instruction

The success of training in overwater mountain river crossing largely depends on skilful selection of the training sites.

Prior to training, set up rescue stations downstream not only on the bank but also in mid-river on separate large stones.

The instructor personally directs the crossing from the bank while located between those performing the exercise area and the rescue station.

The instructor needs to make sure that the students jump lightly from stone to stone and that one arm is free for self-belay. Students should cross with self-belay one-by-one along a stretched primary rope, and they should be on the downstream side of the rope on a sliding carabiner at a leash distance from which they can reach the primary rope.

The two primary ropes should be thoroughly tested when crossing along a rope with a climbing harness and pulley or carabiner. Gear should be well assembled and secured.

To prevent injury and accidents when crossing over water, there should be no:

- training unless the instructor has checked the security and setup of the crossing;
- crossing by the next student until the instructor (mountaineer trainer) checks that the climbing harness fits and that the pulley or carabiner is securely attached to the rope;
- water crossing without belay;
- crossing training before the rescue posts are ready (see Addendum).

Addendum
SAMPLE

"approved"
Commander, Motorized Rifle Battalion
____
(rank, signature, last name)
_______ 2009

LESSON PLAN
Negotiating Mountain Obstacles
_______ Motorized Rifle Company
_______ 2009

SUBJECT:
"Different Ways to Cross Mountain Rivers on Foot"

OBJECTIVES:
1 Train the company in different ways to cross mountain rivers on foot.
2 Develop in company personnel the physical and psychological qualities and endurance to perform various tasks in the mountains.

TRAINING POINTS:
1 Fording a mountain river using different techniques.
2 Water crossing of a mountain river.
3 Moving personnel, weapons and ammunition across a mountain river using ropes and pulleys (carabiners).

TIME: 3 hours

PLACE: Mountain river crossing (forced crossing) training site, mountain training center.

TEXTBOOKS AND GUIDES:
Training Units for Combat Operations in Mountain Regions (Moscow: Voyenizdat, 2007).

UNIFORM:
Summer field gear, with weapons.

SUPPLIES:
Two primary ropes per squad, one backup rope one per squad, one piton hammer per squad, 30 pitons, ten 2-2.5 meter-long wooden poles; one self-belay system, carabiner and pair work gloves per student.

1 Preparatory Part – 10 minutes
Check the training sites (mountain obstacles), personnel attendance, health (by asking), personnel and mountain gear preparedness; state the subject, training objectives and points, training procedure and time at the training sites; review general mountain river crossing requirements; move the platoons to the training sites and perform the preparatory exercises.

2 Main Part – 135 minutes
Training Site No. 1
(45 minutes.)
(1st Motorized Rifle Platoon, Company Command)
1. Select and fit out a ford.
2. Ford a mountain river single file, in ranks, and in a circle.
3. Ford a mountain river along a stretched mountain rope. Instructor: Platoon Commander

Training Site No. 2 (45 minutes.)
(2nd Motorized Rifle Platoon, Flame Thrower and Anti-Tank Squad)
1. Cross a mountain river on exposed stones.
2. Cross a mountain river by log-crossing.
3. Cross a mountain river on a suspendeed pedestrian bridge. Instructor: Platoon Commander

Training Site No. 3 (45 minutes.)
(3rd Motorized Rifle Platoon, Machine Gun and Grenade Launcher Squad)
1. Prepare to cross using a primary rope and pulleys (carabiners).
2. Move weapons and ammunition across using a primary rope and pulleys (carabiners).
3. Move personnel across along two primary ropes on a pulley (carabiner) with self-belay. Instructor: Platoon Commander.

3. Concluding Part – 5 minutes.
Perform muscle relaxation exercises (while moving to the lesson review site); **form** the company, **check** that personnel, weapons and equipment are present; **summarize** the lesson: **determine** how well the techniques for scaling mountain obstacles have been learned, **give grades**, **note** the strengths and weaknesses; **set** tasks for eliminating the weaknesses and preparing for the next lesson.

Company Commander Work Procedure
1. Check the training and crossing sites and organize the rescue station – on the day of the lesson before the company arrives.
2. Check that the company is prepared for the lessons, set tasks – five minutes.

3. Check the progress of the lessons at the training sites and organize belay – 30 minutes. (at the start of the first hour of training).
4. Assist the platoon commanders at training sites Nos. 1 and 3 – 30 minutes at each site (at training site No. 2 – in the third hour of study).
5. Review and summarize the lesson – five minutes.

___Motorized Rifle Company Commander

____

(rank, signature, last name)

Lesson procedure

The training in different ways to cross mountain rivers typically begins with preparatory exercises in an old (dry) riverbed or at a specially equipped training site.

Prior to the mountain river negotiation (forced crossing) training, three sites are fitted out with different crossings.

Prior to the training, the instructor rechecks the river depth and current velocity to determine fordability and organize belay. A rescue station is set up downstream for student safety and belay. In addition, there is a rope stretched by each crossing (at each training site).

When the company arrives at the training area, the company commander states the subject, training objectives, and the procedure for practicing the training points at the training sites, reminds the personnel of the safety measures, and takes the platoons to the training sites.

When the training starts, the company commander monitors the platoon commanders' teaching process, checks the rescue stations, makes sure that the safety requirements are followed, and helps the platoon commanders at training sites Nos. 1 and 3.

At training site No. 1, the platoon commander begins the lesson by briefly explaining and preparing the fording site. As the ford is prepared, individual students practice crossing the river with pole self-belay. At the same time, the end of the rope is taken across to the other bank, the fording is scouted out, and the ends of the rope are secured to the river banks using pitons, crags, trees, etc.

After selecting and preparing a ford, the platoon commander explains and demonstrates on dry land how to ford a river single file, in ranks, and in a circle. He then moves on to practical training on the river, practicing each technique in sequence with the squads, which initially perform the exercises without weapons or marching gear, putting boots on their bare feet, and then in full gear.

As the students master a technique, the platoon commander intensifies the training and makes the squads compete for speed and quality while observing the safety requirements.

Moving on to fording (crossing) a river along a stretched rope, the platoon commander reminds the students that this method is used to cross high-velocity and relatively deep (up to the waist) rivers. He then reminds them of the safety requirements, demonstrates the technique on dry land and in the river, and moves on to practical training. One squad executes the technique on the river under the platoon commander's direction, and the others practices on the bank under the squad commanders' direction.

The platoon commander must be on the bank between the river being forded and the rescue station and personally direct the crossing. He makes sure that the leash with the Prusik knot (carabiner) is of a length that allows a student to reach the primary rope if he is pulled away by the current, and that the students in the water do not raise their legs high or move in step, and that they self-belay.

After all the training points have been practiced, the platoon commander arranges for the uniforms and footgear to be dried (change of clothes and footwear), then gives a brief review and, on the company commander's command, leads the platoon to another training site.

The training at site No. 2 should be conducted in the following sequence. First, the platoon commander explains how to cross a river on exposed rocks, by log-crossing and on a suspended pedestrian bridge. He then checks the crossing and belay ropes, demonstrates each method of crossing a river, reminds students of the safety requirements, and organizes the training by squad.

Each squad practices one of the crossing methods under the direction of the squad commanders, then they change places on the platoon commander's command.

The platoon commander watches the squad training to ensure that the students jump lightly on the rocks, using the pole for support, and that they move slowly and out of step, two to three meters apart, when crossing by log or hanging bridge.

After all the scheduled overwater crossing methods have been practiced, the platoon commander does a review, summarizes the results of the competition between the squads, sets assignments to improve the skills learned, and takes the platoon to the next training site.

At site No. 3, the platoon commander explains how to prepare the crossing with primary ropes and pulleys (carabiners) then assigns students to work on this bank and on the opposite one, and to carry the primary ropes to the opposite bank by fording the river with a pole. Two crossings are then prepared: one for training to carry weapons and ammunition to the other bank, and the other for the students to cross the river. When he has made sure that the platoon knows how to prepare the crossings, the platoon commander moves on to practical training. Weapons (ammunition, gear) are transported first, and then the students cross. Each student should cross to the opposite bank one-two times along a rope on a pulley (carabiner) with self-belay.

The students return from the opposite bank by a suspended foot bridge or by log-crossing.

The commander does a brief review once all the scheduled crossing methods have been practiced.

The platoons change training sites every 45 minutes, on the company commander's command.

The company commander ends the training by critiquing the exercises, noting to what extent the objectives were met, and summarizes the results of the competition between the squads and platoons. In conclusion, he grades the units and gives orders to fix the weak points.

17

Train to Win: Organizing and Conducting Physical Training For Soldiers in the Mountains

---

Colonel Mikhail Panov in *Armeisky Sbornik*, May 2007*

Begin to develop and improve a soldier's physical, special and psychological qualities for com-bat under special conditions before they attend mountain training centers. Use all forms of physical training for soldiers and sergeants to achieve this: morning exercise, classes, physical training during mock combat, athletic activity. Let us look at each of these forms of physical training individually.

Morning exercise should include short and medium distance running, one to five kilometer cross-country races, as well as exercises on multi-purpose gymnastic apparatus (hip pull-overs, pull-ups, arm walking on parallel bars, parallel bar dips), gymnastic bench exercises, and rope and pole climbing. Soldiers should negotiate several obstacle and training areas multiple times or completely negotiate both a general and a mountain obstacle course. Sports with simplified rules can also be included (handball, basketball, soccer), and in summer—going for a dip and swimming.

Conduct **physical training classes** in speed walking, track and field, obstacle courses, hand-to-hand combat, swimming (in summer), and ski training (in winter). In addition, conduct integrated gymnastic classes that include physical exercises, athletic games, mountain obstacle courses, hand-to-hand combat, and others.

After the soldiers have learned to negotiate mountain obstacle courses in the initial classes, they should train to negotiate them with weapons and in gear. Simulate enemy fire and everything typical of mountain combat.

* Colonel Mikhail Panov, "Trenirovat'cya shtobypobezhdat': Ob organizatsii i provedenii fizicheskoy podgotovki voennoslyzhashchikh v gorakh" ("Train to Win: Organizing and Conducting Physical Training For Soldiers in the Mountains"), in *Armeisky Sbornik* (*Army Digest*), May 2007, 46-48.

Collateral **physical conditioning exercises** include natural mountain and water obstacles, speed walking, field hikes, ski treks, etc.

Mass sporting activity should focus on applied sports (general and mountain obstacle courses, military triathlon, track and field, skiing, swimming, etc.) and military-professional techniques and actions.

Before a unit goes to a mountain training center, all personnel undergo a medical exam to see if they are fit to train under special conditions and physical strain. If a serviceman's overall grade is unsatisfactory, he is not allowed to participate in the mountain training but is given additional physical training that focuses on developing endurance and those physical qualities that are poorly developed.

If personnel are well prepared physically on the plains, they will more quickly adapt to activity in mountain conditions, the negative "shifts" in body functioning are mitigated, and the onset of altitude sickness is slowed. Mountain physical training should therefore aim to develop and improve the same physical, psychological, and special qualities and motor skills that are present prior to going to the mountains.

During the first two weeks at the mountain training center, physical training classes are held every day except on days off, and morning exercises take place daily for at least 40 minutes. It is important that there should be not only weight-lifting, body-building and gymnastic exercises but also short, medium, and long distance runs; cross-country races; six to ten kilometer full-field marches; and treks to the mountain training sites (the sites should be 15 to 20 kilometers from the training center). Soldiers must be trained in hand-to-hand combat in mountain conditions, and to throw a grenade accurately and far while standing, kneeling, lying. Also, they must be able to throw a grenade from a trench or bunker, on a level area, as well as downhill and uphill.

In winter, classes should also include ski training (how to move on skis across rugged terrain, negotiating various mountain obstacles on skis, Alpine skiing).

Classes are not organized by physical training category but rather integrate exercises from the various categories. The preparatory part of classes with soldiers who have been in the mountains for less than a week should be increased up to 15 minutes. Their physical load during this period should be light.

There should be two-minute breaks between exercises to catch one's breath and control the pulse rate. As the soldiers adapt (10-12 days after arriving in the mountains), training intensity and load are increased to what they were on the plains. Classes are conducted in accordance with the physical training and sport field manual.

The knowledge and skills that the students gain in the special mountain obstacle classes should be developed in the physical training classes. **Mountain and physical training classes are generally integrated**. This makes possible all-round general physical conditioning of the students, strengthens their health and makes the body more resistant to the impact of high altitude and combat.

Success in mountain combat tasks will largely depend on the students' all-round physical training. Mountain obstacle training should therefore be planned so as to develop all the physical and special qualities of mountain marksmen, where **strength** is a critical quality of mountain combat. And strength is best developed by resistance exercises (barbells, weights, rocks) as well as by pull-ups on a crossbar or other contrivance, rope climbing, etc.

Sample exercises to develop strength:

1. Push-ups on the ground.
2. Two-leg squats without a partner, with a partner, and with weights.
3. One-leg squats.
4. Jumping in place on one leg or both legs.
5. Forward/backward tilt without a partner, with a partner, and with weights.
6. Tugging each other while standing and while sitting.
7. Throwing rocks of different weights with both arms forward and upward, and backward and upward over the head.
8. Pushing stones with one and both arms.
9. Pull-ups on a crossbar, tree, pole, and so on.
10. Dips on bars, poles, overhangs, etc.
11. Arm swings on parallel bars.

In the mountains, the soldiers must be able to relax their muscles so that when some muscles are being exerted, others are resting. Alternately contracting and relaxing the muscles makes it possible to exercise for a long time.

Sample muscle relaxation exercises:

1. Jiggle relaxed arm and leg muscles (one by one).
2. Alternate the body weight between the left and right foot.
3. Lightly massage the neck, chest, stomach and back.

A critical quality for soldiers in the mountains is **endurance**, which depends on the state and efficiency of all human organs and systems. There is general and special endurance: the latter is the ability to negotiate mountain obstacles over a long period of time and march for a long duration and in different terrain.

Walking and running are necessary to develop overall endurance. Best endurance training comes from marches, cross-country races, ski races, and man-made and various natural obstacle courses.

The ability to **maintain balance** is vital for soldiers in the mountains. Without balance, it is virtually impossible to master rock climbing, mountain-river crossing, moving across ice and talus, or conducting belay and self-belay.

Sample exercises for strengthening balance:

1. Balance on one leg.
2. Tilt quickly forward, backward and sideways with eyes closed.
3. Rotate the head and body.
4. Walk and run with and without weapons on a narrow surface.
5. Jump from stone to stone (circle to circle), making a firm landing each
6. time.

Negotiate a multi-event mountain obstacle course.

Swiftness and agility are critical for negotiating the mountain obstacles and dangers related to rockfalls and snow and ice avalanches. Exercises to develop swiftness and agility:

1. Short distance run (100 meters).
2. Running and standing long jumps.
3. Precision throwing of grenades, rocks, and other articles.
4. Negotiate standard and mountain obstacle courses.
5. Perform pommel horse jumps and vaults.
6. Perform different exercises on gymnastics apparatuses and calisthenics.

These exercises are used in all forms of physical training and during mountain training.

During mass sporting activity and rock climbing training, contests should be held to ease the soldiers' and sergeants' emotional strain.

The final stage of training should be a graded ascent to a peak or pass, and a qualifying test for the right to the title and badge of "Alpinist of Russia."

The following exercises are recommended to test the physical training done at 2,500 meters above sea level:

One kilometer cross-country run in military uniform and hiking boots to measure endurance ("excellent"—4 minutes 30 seconds, "good"—4 minutes 50 seconds, "satisfactory"—5 minutes 30 seconds);

Walk along a narrow ledge (length—six meters, breadth—20 centimeters, height above the ground—two meters) to gauge balance ("excellent"—walk quickly and confidently, "good"—walk without stopping, "satisfactory"—walk with stops and without falling off); **jump from a height of two meters onto a 50x50 centimeter square on the ground** to gauge precision and orientation in the air ("excellent" – land on the square and stop without hands touching [the ground] and no forward or backward step, "good"—land slightly outside the square, "satisfactory" – same, but with hands touching the ground); **rope climbing (primary rope, doubled) without using the legs** to gauge arm strength ("excellent"—seven meters, "good"—six meters, "satisfactory" – five meters); **one-legged squat** (on an even platform, bench or rock; the other leg is lowered freely or stretched forward; arms are held in an optional position; the knee of the squatting leg touches the chest with each squat) to gauge leg strength ("excellent"—12 times, "good"—10 times, "satisfactory"—8 times); **15 meter jumping race on stones** (five meter race, then jump from rock to rock; rock height – at least 10 centimeters; nine-ten rocks should be spaced over 10 meters in such a way that either the right or the left foot can tread on a rock once; with the last jump, land on both feet, stopping in the designated square on the finish line) to gauge reaction and motor coordination ("excellent"—3.9 seconds, "good"—4.0 seconds, "satisfactory"—4.5 seconds).

Physical fitness is tested over the course of one day in the following sequence: walk along a narrow surface, jump from a height, rope climbing (primary rope), one-legged squat, jumping race on stones, one-kilometer cross-country race.

The physical fitness grade is made up of the grades received for the six exercises, namely:

"**Excellent**" – if at least half of the exercises are graded "excellent," and the rest "good";

"**Good**" – if at least half of the exercises are graded "good" or "excellent," and the rest "satisfactory";

"**Satisfactory**" – if more than half the exercises are graded "satisfactory" and there are no unsatisfactory grades.

A soldier's fitness is graded "**unsatisfactory**" if there is even one unsatisfactory grade.

Build and equip sites and sport facilities for personnel physical training.

Sport facilities are built and sites equipped for military units assigned to mountain operations in accordance with the "Suggested List of Sport Facilities and Sites for Physical Training" (Addendum 2 to the *Handbook of Physical Training and Sport in the Russian Federation Armed Forces*). Special mountain training camps or training sites are also built.

Mountain training centers usually have track and field grounds, gymnastic courts with a gymnastic floor and multipurpose gymnastic equipment; rings for hand-to-hand combat; an area for throwing grenades accurately and for distance, downward and upward; sport courts (for volleyball, basketball, soccer, handball); and, if conditions allow, an outdoor swimming pool or aquatic facility.

LOGISTICS

Logistics in the Mountains

S*upply in the mountains is tough. Helicopters cannot reach many areas, trucks have a hard time negotiating mountain roads, and diesel engines have real problems at altitude. River transport is one answer to resupply through the mountains and, though tricky, rivers have been used in this role for centuries. The following two articles by Vladimir Shiryayev discuss the need to move supplies by river, pack animal, and truck as well as do a logistics water reconnaissance. A mountaineering instructor with much experience in military and adventure rafting, Shiryayev presents ideas that have a great deal of value in mountain combat and are being considered for adoption by the Russian military.*

The difficult mountain rivers of Afghanistan make it hard to move large amounts of material and the times they can be used in this role is seasonable. Still, serious mountain combat requires the ability to penetrate beyond the first mountain range and often pack animals, water transport and porters can penetrate where trucks and helicopters cannot.

18

Hydro-Meteorological Support of Military Transport Along Mountain Rivers

---

Vladimir Shiryayev in *Armeisky Sbornik*, September 2009*

Using mountain rivers for supply deliveries, evacuation of the wounded and sick, and other military transport missions require reliable hydro-meteorological support since they are completely dependent on the hydrographic network in combat areas. Transport operations in such unusual and nontraditional conditions will be successful if there is good hydrologic reconnaissance, which is the basis of hydro-meteorological support for rafting.

Since the objective of hydrologic reconnaissance is to gather, process and provide information to the troops about the hydrologic network and to predict probable natural phenomena in mountain river basins, it needs to be provided by special military support. It is a key component of rear-area reconnaissance since it determines the routes for delivery and evacuation using mountain rivers, and makes transport safety recommendations. Given the many matters that hydrologic reconnaissance deals with, it may be conducted by the staff of large formations as well as subordinate logistic support entities that make deliveries to their own units and detachments within a certain segment of the logistics route.

The hydrologic state of a riverine network can be studied while preparing for and conducting an operation. The study may be based on the following key objectives:

- ascertain the hydrologic and topographic conditions in order to select a staging area;

* Vladimir Shiryayev, "Gidrometeorologicheskoe obespechenie voinskikh perevozok, osyshchestvlyaemikh po gornim rekam" ("Hydro-Meteorological Support of Military Transport Along Mountain Rivers"), in *Armeisky Sbornik* (*Army Digest*), September 2009, 15-22. Shiryayev is a military mountaineering instructor in Tashkent, Uzbekistan.

- study the shoreline to determine the capacity of the vortexes (eddies) for simultaneous loading of watercraft in the staging area and unloading of supplies at the destinations, or for transferring supplies at places where they must be portaged around impassable obstacles during stage-by-stage and segment transport [in stage-by-stage transport, the transport unit moves the entire length of the river, moving supplies from one dump site to the next in states. In segment transport, the transport unit moves the supplies along one segment of the river and then another transport unit moves them further];
- search the riverside in a valley (gorge) for suitable and secure locations for field depots and supply dumps;
- inspect dangerous rapids, narrow passages, and forest obstructions, and make recommendations on how to negotiate them;
- determine the suitability of mountain river stretches for sending supplies by unmanned raft, the forces and assets required for those purposes, and the types of rafts and packaging;
- determine the work to clear the riverbed of minefields, nonexplosive obstacles, and materiel and gear that have fallen into the water;
- determine conveyance methods based on route difficulty and rafting conditions;
- work up transport safety recommendations;
- mark water conveyance routes and set up beacons and directional signs;
- determine the possibility of safe conveyance and list appropriate actions in support;
- prepare charts and technical descriptions of individual obstacles as well as the water routes as a whole;
- do a time study of the rafting, including surmounting difficult-totraverse obstacles (individually);
- inspect the shoreline and riverside in the staging area and at destinations and intermediate points to identify minefields, loading and offloading conditions, and the possibility of using machinery for loading and offloading supplies from undeveloped banks;
- work up proposals and recommendations for camouflage, communication (signaling), engineer and other types of combat support of transport, and for the possibility of nighttime transport.

The scope and complex nature of hydrologic reconnaissance dictates the need to form reconnaissance groups from different services and specialties: rear services, hydro-meteorologists, engineers, com- munications, and chemical defense specialists. There could also be a need for hydrologists; glaciologists; camouflage specialists; demolitions experts; fork-lift transport vehicles and machinery experts; field fortification experts; and sappers. The size and composition of hydrologic reconnaissance groups are not fixed and will vary widely depending on the scope and nature of the information required.

The chief methods of obtaining information about a river are: large scale topographic maps, global positioning system readings; instrument and visual observation; ground and aerial photography; video recordings; and direct inspection of obstacles, shore lines and riversides, paying particular attention to planned intermediate mooring sites for stage-by-stage and segment conveyance. It is also necessary to thoroughly inspect the riverside in the staging area and at the supply destinations, whatever the means of transport.

When gathering information on a river, it is vital to familiarize oneself with its technical description and charts, reference and special literature, and the military-geographic description of the theater of operations. The author believes that it would be a good idea to print large-scale topographic maps with river characteristics annotated on them. The data would be derived from trekking reports. These brief characteristics of rafting conditions would be a great addition to the written information on the reverse side of the map.

Charts make transport much easier, quicker and safer by making it possible to take one's bearings in a river setting and make appropriate decisions. The charts enable detachments to move and make supply deliveries more rapidly. Figure 1 shows navigational directions for a section of a mountain river that the author drew up for a difficult trek.

Charts and technical descriptions compiled back in peacetime can certainly be used for military transport, but the information they contain could be outdated by changing terrain factors (landslides, avalanches, rock falls, and so on) and the impact of manmade activities on the environment. Charts could become rapidly outdated in wartime when heavy use of artillery and aviation, and of armored formations and units in areas accessible to tanks, cause a certain degree of terrain changes in mountain river valleys, on hill sides as well as in the riverbeds themselves. Furthermore, explosions and blasts could result in the appearance of new micro-relief features—rapids, shallow rocky rapids, waterfalls, and rock dams, and where there is snow cover and in high mountains—snow bridges and tunnels caused by rock falls and snow slides. The latter could sweep debris in its wake, often reaching the opposite slopes of a valley (gorge). Such obstacles can also appear in mountain rivers during fortification of strongholds, firing positions, command posts, and supply dumps. These terrain changes cannot, of course, obliterate water routes, but by making rafting conditions more difficult, they will have direct impact on water craft and transport.

Charts play a substantial role for military transport using mountain rivers because they provide essential information on rafting conditions and guide selection of tactical rafting options. An expertise in chart reading and chart drawing is a key component of water assault training.

Experience shows that when drawing a chart, it is a good idea to trace the river line and its adjacent terrain from a topographic map with a scale of 1:100, 000 or 1:50,000. For correct orientation when rafting, the tracing must be marked with the mouths and names of rivers and creeks that join the water route, elevations, canyons,

Navigability	Corresponds to River Class (Section)	Current Character	Current Speed, meters/sec
Easy Navigable	One	Still	Up to 2
Moderately Navigable	Two	Rapid	2-4
Difficult to Navigate	Three	Stormy	4-6
Non-Navigable	Four	Very Stormy	Over 6

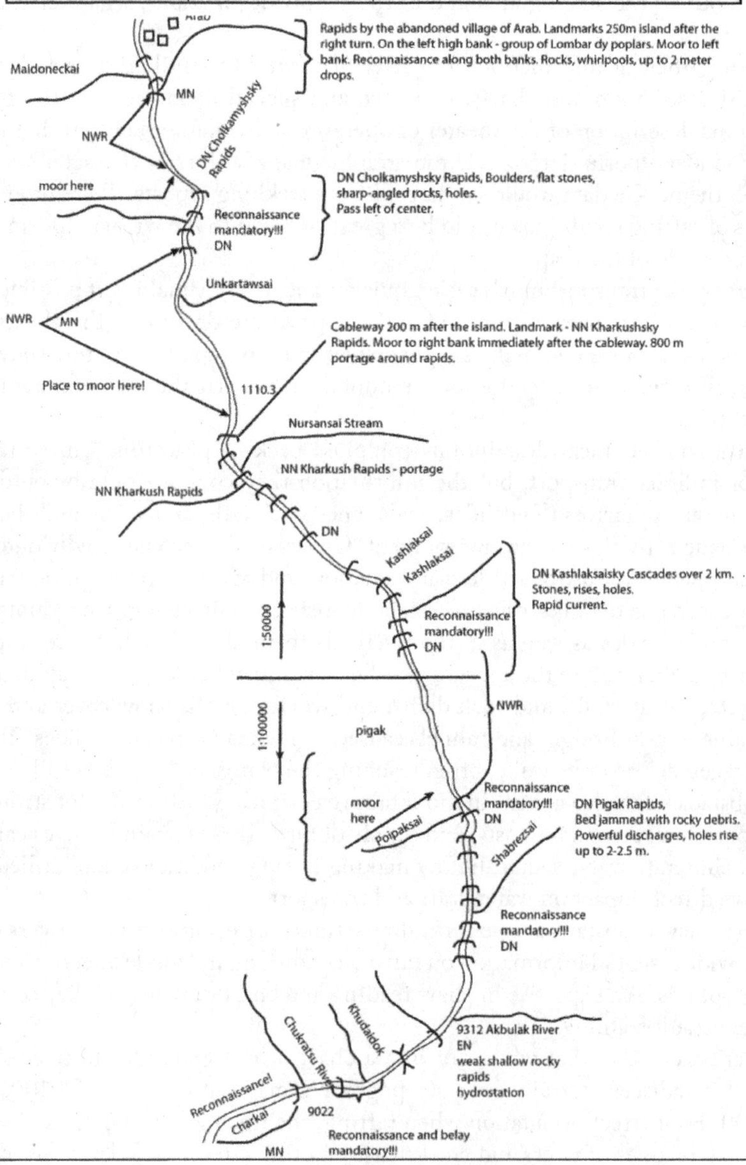

Figure 1 Chart of a mountain river stretch.

bluffs, waterfalls, hydro- stations, cableways, ferry sites, and trails running alongside the river (figure 1).

During hydrologic reconnaissance, all obstacles and their characteristics, and the recommended methods for clearing them are promptly marked on the traced river line. Information should not be entered after reconnaissance as this could result in distortion. Rafters may mark the information immediately on the mapsheet, but it may end up being very crowded because of the large volume of river data.

The diagrams should use a legend, for which the author offers some suggestions (figure 2), and it is a good idea to note the obstacle name (number), location and difficulty as appropriate, for example, 2EN, 5DN, 9NN, and 17MN. Here, the numbers "2, 5, 9, and 17" indicate the sequence number of the obstacle from the staging area,

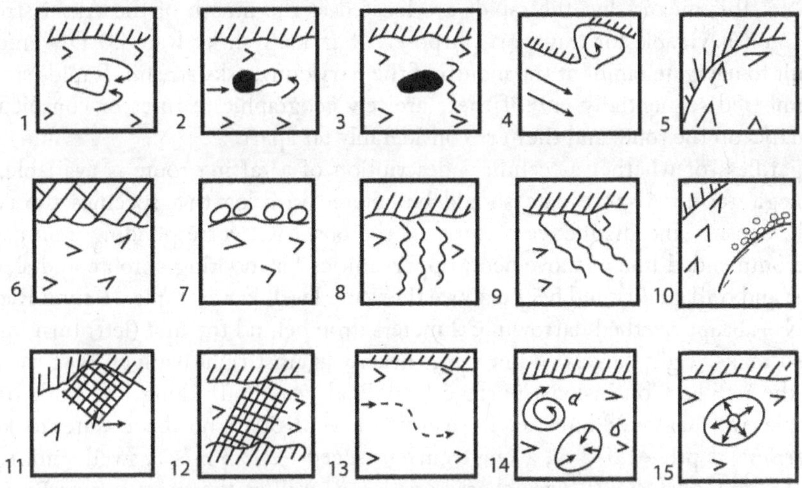

Figure 2. Examples of legends for charted obstacles

1. Above-water rock.
2. Rock partially submerged at a shallow depth.
3. Submerged rock that could impact craft position if traversed.
4. Eddy (section of a channel with reverse current).
5. "Pin spots" where the current may pin rafts against the shore (forms at a concave bank).
6. Rock (bluff bank).
7. Riverside rocks (rock debris, detritus).
8. Vertical rises.
9. Slanting rises.
10. Riverside shoal (forms at a convex bank).
11. Log barrier (riverside).
12. Log barrier (solid).
13. Navigable channel on some river stretches.
14a. Whirlpool.
14b. Swirl.
15. "Mushroom" water cushion, usually pulsating.

Recommended legend for charting navigability

EN – Easily navigable.
MN – Moderately navigable.
DN – Difficult to navigate.
NN – Non-navigable.
NWR – Navigable without reconnaissance but reconnaissance would be useful.

Figure 2 Examples of legends for charted obstacles

and "EN, DN, NN and MN" indicate the degree of difficulty. In this instance, the second obstacle is easily navigable, the fifth is difficult to navigate, the ninth is non-navigable, and the seventh is moderately navigable.

Remember that on mountain rivers which might flood after rain or which have some form of controlled flow there could be significant daily fluctuations in water level, causing obstacles to be partially submerged or to appear in other places where current charts indicate that there are none. Obstacles can also emerge when a mud flow carries debris from lateral gorges (at both low and high altitudes) during glacial mudflows.

Therefore, the numbering of rapids marked on old charts should be checked against the location of lateral flows. And if an obstacle is by the mouth of a lateral flow, it is a good idea to add a geographic name to the numbering, for example, "Aksai Rap. 25 NN." This means that the rapids are located at the mouth of the Aksai stream and is non-navigable for transport purposes. "Sarykamyshsky Rap. 20 DN" means difficult to navigate rapids at the mouth of the Sarykamyshsky stream. Rapids should be numbered sequentially only if there are few geographic features or conspicuous landmarks on the route and they are considerably far apart.

Regardless of whether a technical description of a rafting route is available, all hydrologic reconnaissance data should be entered in a log that specifies the type, length, density, and distinctive features of the obstacles; corresponding landmarks; the recommended line of movement; opportunities for mooring enroute and size of eddies; and trail quality and height above the water level. For example, "Krutoi Rapids 27 DN—abrupt riverbed narrowing; 2 meters drop behind the first (left) turn; many flooded rocks, 'tight barrels'; water slams heavily against right bank cliffs at the exit from the rapids. Conclusions: portage loads and watercraft along left bank trail." Example two: "Calm 28 km stretch, river width—40–50 m, no above-water rocks or log barriers, depth—1.5–2 m, average current speed—2–2.3 m/sec, swell—not more than 0.5 m. Conclusions: these characteristics, as well as the smooth turns, no pin spots or rocky banks make possible through conveyance in either piloted vessels or by unmanned raft."

Completeness should be a consideration when assessing the quality of mountain trails that might be used for stage-by-stage and segment conveyance. It often happens that these primitive transportation lines marked on a map are virtually nonexistent in some stretches. Landslides, mudslides, runoff on slopes, and grass and shrub overgrowth gradually remove these land supply lines. The sharp drop in tourists in mountainous areas over the last 16 years has had a negative impact on trails and their proper maintenance.

If time allows, it is a good idea to diagram rapids on a 1:1000 scale, marking only those rocks, cliffs, currents and whirlpools that could affect passage. A diagram of rapids should show the recommended line of movement for vessels and mooring sites for offloading or transferring supplies to pack loads if the river route cannot be followed further. Trails on which the load and watercraft can be portaged past the rapids to continue the trip should be indicated.The broken line in Figure 3 shows the safest rafting route in very difficult rapids. The vessel in the right turn of the river moves rightward early to avoid being pinned against left bank cliff at 3. The steeper river gradient drastically increases current speed from location 2. In order to avoid eddy 4 when going from one end of the rapids to another, the crew rows vigorously to the left immediately after pin spot 3 and keeps as close as possible to the left bank, avoiding the tangle of submerged and flooded rocks 5. Making allowance for the fact

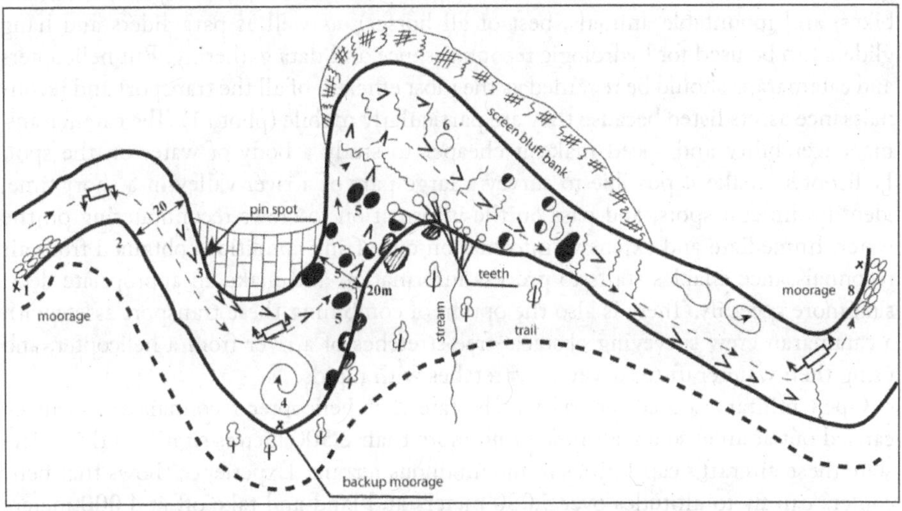

Figure 3.

Photo 1 Bird's eye view of a mountain river's channel and flood plain.

that the river again makes a right turn and that the water slams heavily against the concave (left) bank, and that there is the danger of the craft striking above-water cliffs 6, and then 7, which are aligned, the crew makes every effort to go rightward and passes close to the right bank. Having cleared the final stretch, the raft crosses the current and moors at sandbar 8 on the left (convex) bank.

Watercraft (catamarans, inflatable boats, flotation rafts and kayaks), helicopters, ground vehicles (motor vehicles, armored personnel carriers, motorcycles, mountain bikes, and mountable animals, best of all horses), as well as paragliders and hang gliders can be used for hydrologic reconnaissance and data gathering. But helicopters and catamarans should be regarded as the most efficient of all the transport and reconnaissance assets listed because they are particularly mobile (photo 1). The catamarans' maneuverability and speed make it cheaper to study a body of water on the spot. Helicopters make it possible to survey a large part of a river valley in a short time, identify difficult spots, and pass on the information to crews reconnoitering on the water. Immediate and extensive information on rafting conditions obtained from air reconnaissance enables staffs to process information and make an appropriate decision more speedily. There is also the option of combining these transport assets with a catamaran crew surveying obstacle-free stretches of a river from a helicopter, and using their watercraft to survey the stretches with rapids.

Open military source reports which state that helicopter reconnaissance can be carried out at an absolute altitude of no more than 2,500 meters significantly understate these aircraft's capabilities in mountainous terrain. Experience shows that helicopters can fly to altitudes over 5,000 meters and land and take off at 4,000 meters altitude in hard-to-reach areas. For example, MI-4 helicopters delivered supplies (construction materials, equipment, diesel engines and other machinery) to an altitude of 3,800 meters above sea level during construction of the Lednik Abramova (Abramov Glacier) high mountain hydro-meteorologic station. Furthermore, the tactical radius for deliveries was 50 kilometers, and the cargo weight was 0.7–0.8 tons. Numerous drops to mountain climbing expeditions in Central Tyan-Shan and Pamir, as well as glaciologic research in other high mountain areas of Central Asia confirm the considerable high altitude capabilities of the helicopters (MI-8MTV).

Hydrologic reconnaissance would be incomplete if it were confined to gathering information about the water transport route and the riverside area where supply dumps are to be set up, supplies loaded and offloaded, and loads and watercraft portaged—for segment and step-by-step transport. Hydrologic reconnaissance is especially important for enhancing the ability of troops and logistic services units to survive in high mountain areas and must thoroughly and regularly survey the man-made (dammed) and glacial (ice-dammed) lakes that abound in the basins of rivers with glaciers and a lot of erosion terrain formation (photo 2). Some glacial lakes break out regularly. The break-out of the well-known Mertsbakher ice-dammed lake in Central Tyan-Shan can, to some extent, be forecast since it takes place annually (from late July to mid-October, and most often in August-September) causing powerful floods on the Inylchek and Sarydzhaz Rivers with a peak discharge of 1,000–2,000 cubic meters per second. Still, it is impossible to forecast the break out of the enormous number of small lakes that form on the surface and in the moraine deposits of glaciers.

Consequently, helicopter survey of the high mountain area of river basins is important where there is combat and logistics support activity. It makes sense to land observers with radios and safety and survival gear at the most potentially dangerous high mountain lakes to study and analyze the processes underway at the site, record the level of the water accumulating on the surface and in the moraine deposits of glaciers, as well as to preempt attempts by the enemy's sabotage and reconnaissance groups to blow up the glacial (rock) dams to deliberately flood the locality.

As regards the need for aerial hydrological reconnaissance of high mountains (snowcap region), mention should be made of the SARNIGMI hydrographic party's many years of experience conducting such observations back in Soviet times. Unfortunately, the economic hardships that set in after 1991 sharply reduced the scope of aerial and ground hydro-meteorologic observations. The disastrous July 1998 glacial mudflow in the Shakhimardan River valley resulted from the melting of glacial dams, followed by water discharge from a small lake that had formed in the terminal moraine deposits of the ArchaBashi glacier (Alai Mountain Range, photo 3). This vividly disclosed the severity of the current challenge of ensuring the safety of the whole area.

Later overflights of the Pamir-Alai and Western Tyan-Shan mountain ranges identified lakes whose parameters were determined visually: length, width, shape, surface area, shore line configuration, and approximate water volume. The surface of each glacial basin was surveyed. When a body of water was found, the helicopter would pass over the site several times until all information, including the exact time and location, were logged. Relevant notes were also made on topographic maps. Special

Photo 2 Erosion terrain formation.

Photo 3 Glacial mudflow.

reference books and catalogs of USSR glaciers were used in the aerial survey. Republic of Uzbekistan Ministry of Emergency Situations hydro-meteorologic service detachments landed on lakes that were potentially dangerous.

Lakes that emerge from the ice twitches [the sudden downstream movement of the ice cover of a glacier] of glaciers formed in lateral gorges to the river can block the principal riverbeds and present a major danger to troops and logistics units. Some glacial dams can be 200 meters high. If they burst, they will cause catastrophic flooding of mountain rivers, destroying transport supply lines and towns. The Voyenno-Gruzinsky road has, on more than one occasion, been put out of commission by mudrock flows on the Kabakhi River and then on the Terek River. Transport supply lines in Alaska, including Stikine River traffic, are currently threatened by at least six glacial lakes. The torrential flooding that occurred on the Rio Plomo and then on the Mendoza River in January 1934 wiped out seven bridges and 13 kilometers of the Trans-Andean Railway along the Mendoza gorge and through the Cumbre Tunnel from Argentina to Chile. It took 10 years to rebuild the railway. In Karakorum, the world's largest mountain range, there have been at least 50 glacial lake bursts over the last 150 years, spawning colossal floods. The water level in some rivers rose by more than 20 meters, and the discharge reached 19,200 cubic meters per second with a water flow of 30–40 kilometers per hour. During one such flood on the Shyok River, a large right tributary of the Indus, villages on the lower terraces were damaged while those high (60–90 meters) above the river were not affected by the flood wave, even though it reached 15 meters (Yu.B. Vinogradov. *"Etyudy o selevykh potokakh"* ("Sill Flow Etudes"), L., 1980).

These phenomena and the scale of the destruction they wreaked show how vital it is to ensure the safety of the troops and logistics units and agencies, to which end personnel, weapons, vehicles, supplies, and command posts should be located out of the reach of flood waves and sill flows.

Hydrologic reconnaissance groups may use the suggested classification of natural obstacles by navigability and the table data to assess a water surface and issue recommendations for using mountain rivers as military supply lines (transport and evacuation routes).

Easily navigable obstacles. They may take the form of single rocks, several sandbars, small waves, weak pin spots, straight slightly sloping flows and rapid flows (photo 4).

All types of watercraft with maximum supply loads may be used with such obstacles. The vessels' line of movement is determined by reading the water. Belay can be used. The vessels could also be joined onto more powerful watercraft or separate sections of a pontoon bridge could be used as unmanned rafts to be intercepted later at the offloading site. Positive velocity (strokes with oars forward) can be used to increase speed on rowed rafts, and especially on oar-propelled vessels (catamarans, flotation rafts, and inflatable boats). Stretches with easily navigable obstacles are convenient for refining tech- nical and tactical rafting techniques such as spacing between vessels and maintaining contact, crew belay, reading the water to determine the line of movement and following it.

Moderately navigable obstacles. They allow a trained crew to move loaded watercraft. This obstacle category includes simple rapids and shallow rocky rapids, some rock groups, slight rises, stronger pin spots and rapid flows (photo 5). These obstacles can be traversed by all types of watercraft with belay. In some cases there may be a slight drop in speed when obstacles are re- connoitered from shore and belay is organized from the water. The commander of the rafting detachments determines the distance between crews based on the situ- ation and the character of the river. Rafting experience shows that vessels should be spaced approximately 30–100 meters apart to avoid collisions while moving and ensure crew safety during belay. The vessels' line of movement in most cases is determined by reading the water.

Photo 4 Easily navigable obstacles.

Photo 5 Moderately navigable water obstacles.

Difficult to navigate obstacles. These are characterized by rapid currents and great difficulty – powerful rapids and pin spots, slanting and pulsating rises with reversed crests, foamy holes ("kettles," "barrels"), powerful whirlpools, tight vortexes, current disturbances, swirls, abrupt discharges, and rocky pockets. It takes a lot of courage to clear these obstacles. If they follow one another in quick succession, plan on making the trip in vessels that have a lightened load or have been completely unloaded (photo 6). The distance between the crews should be increased but should not exceed 100 meters. In some cases, belay is completely eliminated during movement and each crew is on its own. Load and watercraft portage is not ruled out. Obstacles in this category can be successfully cleared in maneuverable vessels that are stable and fairly buoyant. If there are enough forces

Photo 6 Difficult-to-navigate obstacles.

Photo 7 Non-navigable obstacles.

and assets, it is a good idea to plan to belay from the water downstream of the rapids. Difficult-tonavigate obstacles on transport and evacuation routes can significantly slow supply delivery and evacuation of the wounded. In some cases it will take two-three times longer than a trip on easy stretches of the same length, which is based on just travel time. It may take as long to make the first trips on unexplored routes and when there are insufficient forces and assets for safety.

Non-navigable obstacles. It is impossible to clear them on any type of watercraft, even unloaded, or with a highly trained crew (photo 7). Mooring teams must be dispatched beforehand to help crews when such obstacles are present on a conveyance and evacuation route. This is all the more necessary in cases where watercraft mooring is compulsory and unmanned flotation rafts, separate sections of a pontoon bridge and other rafts on the mountain river have to be intercepted. Examples of this group of obstacles are waterfalls, rock and log barriers, and rapids with macro-fragmental jams where narrow passages make rafting impossible. With engineer support, some types of these obstacles, for example solid log barriers, can be removed to provide an unobstructed passage and evacuation route.

The suggested organization of hydro-meteorologic support for military transport on mountain rivers, and the classification of natural obstacles by navigability, make it possible to use water routes very efficiently and safely as military supply lines in hard-to-reach localities. The foregoing recommendations may be useful for the staffs and logistics services of formations and task forces when suddenly called upon to provide uninterrupted supplies to troops, move military contingents, and provide material and transport support for combat and operations in special conditions.

M*ountain logistics is tough and requires ingenuity and some ancient skills to succeed. It is apparent that:*

1. *Mountain artillery support is a function of how far forward artillery ammunition can be delivered.*
2. *155mm artillery systems are hard to get out of valley plains and smaller, direct-fire artillery or rocket launchers that can be broken down into transportable pieces are optimum (75mm or thereabouts).*
3. *Mortars are great, but like aircraft, give the enemy time to react to incoming fire. Mortar ammunition above 81mm is hard to manpack.*
4. *Most soldiers do not know how to care for, load and lead pack animals, but this can be a vital skill.*
5. *Just because its water does not mean it is a job for the US Navy.*
6. *Transportation units need to be capable of self defense.*
7. *Helicopter supply is a limited option.*
8. *Transportation units need mountain and river training just as much as maneuver forces.—Editors*

19

What a Water Transport Company Should Be: The Need for Special Supply Delivery Transport Units in Mountain Areas

Vladimir Shiryayev in *Armeisky Sbornik*, August 2008*

The challenging geography and physical conditions in mountainous regions greatly affects soldiers' combat and all types of logistics support. Material and transport support and the techniques and methods of supply delivery require an improved logistics support system, redundant delivery supply routes, and, most importantly, new specialized transport units.

The training of military units using the proposed TO&Es shows that shows that a separate water transport company (mountain)—SWTC(M) meets all the requirements for mountain operations and is best suited for this purpose. The company should include the prescribed number of water transport platoons and squads, as well as other combat, logistical and technical support units (figure 1).

The suggested mountain water transport company's TO&E structure will meet the daily requirements of a mountain battalion using up to a class three river with severe rafting conditions and a five-fold buoyancy (figure 2). A 38-man water transport platoon is best suited as the mountain water transport company base unit. Comprised of three water transport squads, it has a 52.5 ton maximum carrying capacity (the capacity of all inflatable elements).

The optimum number of men for a water transport squad is twelve. The crew may have two, three, four and sometimes six people, depending on rafting conditions. A squad can deliver a mountain platoon's daily material requirements in one trip.

It makes sense to use the wire-framed inflatable rafts on transverse gondolas and catamarans as the separate mountain water transport company's primary regular

* Vladimir Shiryayev, "Kakoy byt'vodno-transportnoy rote?" ("What a Water Transport Company Should Be"), in *Armeisky Sbornik* (*Army Digest*), August 2008. Shiryayev is a military mountaineering instructor in Tashkent, Uzbekistan.

watercraft. Each has a maximum carrying capacity of 7.5 and 2.5 metric tons, respectively.

Figure 3 shows the carrying capacity and minimum number of recommended watercraft (one set) needed to perform the tasks of a separate mountain. With 4-5-fold buoyancy, it is seemingly inadvisable to use catamarans for cargo transport since their carrying capability is considerably less than rafts.

Counting the weight of the crew and weapons (400 kilograms), their useful carrying capacity is just 230 and 100 kilograms, respectively. But even under such conditions (technically difficult rapids), a mountain water transport company can carry 32.5 metric tons and deliver 20 metric tons of useful cargo to fill a mountain unit's daily material requirements.

NOTE: *The author proposes using this company for both upstream and downstream transport. Upstream transport will require rowing, belaying, or towing from the banks. Where this is impossible, he envisions use of vehicles and pack animals to bridge the gap in moving supplies and disassembled watercraft upstream.*

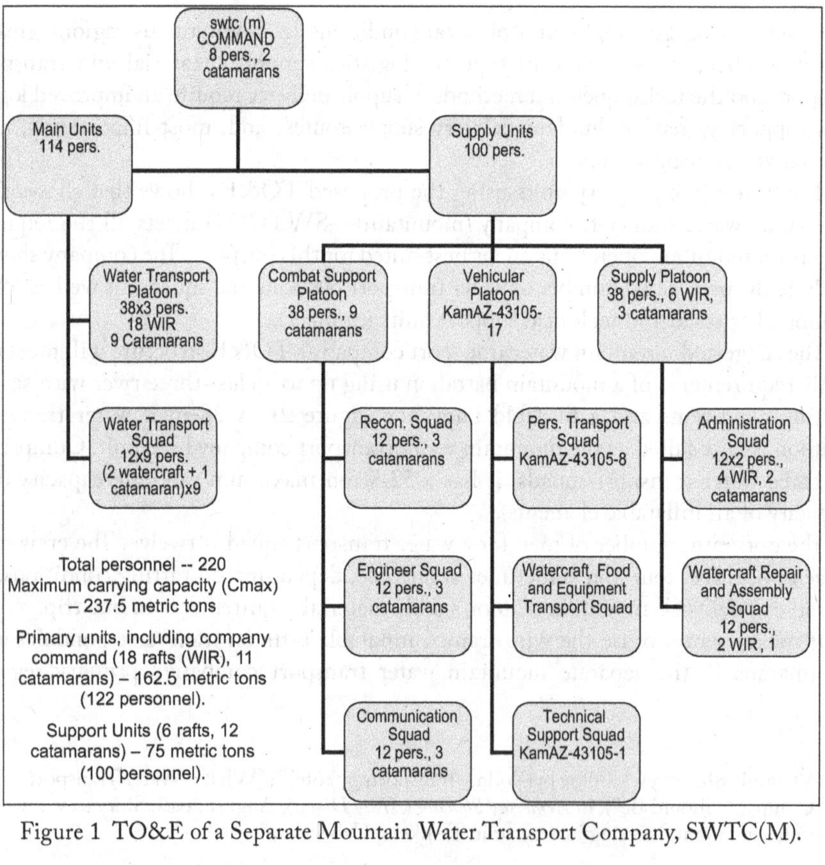

Figure 1 TO&E of a Separate Mountain Water Transport Company, SWTC(M).

River Rafting Classifications and Bouyancy

Class 1: Smooth river with only small rough areas that might require some maneuvering. (Skill Level: Very Basic)

Class 2: Some rough water, perhaps with some rocks that might require some maneuvering. (Skill Level: Basic)

Class 3: Whitewater with small waves and perhaps maybe a small drop, but no real danger. It might require significant maneuvering. (Skill Level: Experienced)

Class 4: Whitewater with medium waves, perhaps with rocks and a considerable drop requiring sharp maneuvering. (Skill Level: Experienced)

Class 5: Whitewater with large waves, fast water, probably including large rocks and hazards and a large drop that requires skilled maneuvering. (Skill Level: Advanced)

Class 6: Not navigable due to substantial whitewater, large waves, rocks and hazards, plus substantial drops that may destroy water craft.

Four or **five-fold buoyancy** refers to a water craft's ability to carry four to five times its weight.

Information extracted from http://en.wikipedia.org/wiki/Rafting assessed on July 15, 2010.

Figure 2.

Water Transport Units	No. of Personnel	No. of Rafts on Gondolas	C_{max}' (Tons)	No. of Catamarans	C_{max}' (Tons)	$\Sigma\Gamma$ max' (Tons) Water Transport Units	Class Rivers Carrying Capacity (when River Class) 3	4	5
Crew	4	1	7.5	1	2.5	7.5:2.5	2.5;0.83	1.88;0.63	1.5;0.5
Squad	12	2	15	1	2.5	17.5	5.83	4.4	3.5
Platoon	38	6	45	3	7.5	52.5	17.5	13.1	10.5
Company Command	8	–	–	2	5.0	5.0	1.67	1.25	1.0
Company	122	18	135	11	27.5	27.5	54.17	40.63	32.5

Total of 29 watercraft in a set: 18 rafts; 11 catamarans.

Figure 3 Carrying capacity and minimum number of watercraft needed for a Separate Water Transportation Company (Mountain), SWTC(M), minus support units.

Inclusion of catamarans in platoons and squads ensures crew safety since they are more reliable for belaying large watercraft and intercepting crews undergoing some emergency.

One of the best separate mountain water transport company manpower distribution options on watercraft is as follows: 44 servicemen, including the company command, carry supplies on catamarans, while 78 carry them on rafts. There are 12 four-man raft crews and the rest are made up of five-six men. This manpower distribution on the rafts has to do with the disposition of platoon commanders and their deputies. There are other possible crew configurations, depending on rafting conditions.

Platoon commanders should be appointed from among officers who have been trained in combat rafting, and their deputies, as well as squad and crew commanders should be appointed from among the chief warrant officers, sergeants and privates who are best trained in mountain river rafting.

During rafting, the platoon and squad movement columns are to be formed in such a way that a catamaran crew is at the head of each unit pointing out the route as well as dangerous local obstructions and, if necessary, berthing sites.

A drawback of this military unit model is that the separate mountain water transport company is outfitted with rigid-frame watercraft, which means that a lot of time and materials are spent making them. In southern regions and the Thule, where there is no forest cover, watercraft frames have to be made from dural (titanium) tubes, making them more expensive.

This drawback is compensated for by the long useful life of watercraft frames made from these metals (ten years or more based on the experience of sport rafting). These watercraft can be disassembled into pieces weighing no more than 8 kilograms each, making it possible for the equipment bearers (the very same rafts-men) to carry them to the staging area (SA) after the trip. Since the catamarans have adjustable crossbars, it is possible to change the width of the watercraft so that they can negotiate the narrowest waterways and to arrange and brace the supplies conveniently and securely.

A separate mountain water transport company should have a second and even third set of watercraft in order to reduce the time between trips. They could also serve as company reserves in the event that craft are lost, and to replace craft that are temporarily out of commission. Additional sets will greatly speed up supply delivery. For example, a second catamaran is prepared to leave while the first is transporting a load.

<table>
<tr><th rowspan="4">Water Transport Unit</th><th colspan="9">Mountain Rivers</th></tr>
<tr><th colspan="4">Class One</th><th colspan="5">Class Two</th></tr>
<tr><th rowspan="2">Unit Composition (Men)</th><th rowspan="2">No. of Rafts</th><th colspan="2">Carrying Capacity (Tons)</th><th rowspan="2">Unit Composition (Men)</th><th rowspan="2">No. of Rafts</th><th colspan="3">Carrying Capacity (Tons)</th></tr>
<tr><th>Max</th><th>Class 1.5</th><th>Max</th><th>Class 2</th><th>Class 3</th></tr>
<tr><td>Crew</td><td>2</td><td>1</td><td>7.5</td><td>5.0</td><td>4</td><td>1</td><td>7.5</td><td>3.75</td><td>2.5</td></tr>
<tr><td>Squad</td><td>12</td><td>6</td><td>45.0</td><td>30.0</td><td>12</td><td>3</td><td>22.5</td><td>11.25</td><td>7.5</td></tr>
<tr><td>Platoon</td><td>38</td><td>18</td><td>135.0</td><td>90.0</td><td>38</td><td>9</td><td>67.5</td><td>33.75</td><td>22.5</td></tr>
<tr><td>Company Command</td><td>8</td><td>4</td><td>30.0</td><td>20.0</td><td>8</td><td>2</td><td>15.0</td><td>7.5</td><td>5.0</td></tr>
<tr><td>Company</td><td>122</td><td>58</td><td>435.0</td><td>290.0</td><td>122</td><td>29</td><td>217.5</td><td>108.75</td><td>72.5</td></tr>
</table>

Figure 4 Second model Separate Water Transport Company (Mountain), SWTC(M), supply carrying capability.

The physical and geographical differences of a mountain region dictate the specific approach to the structure, composition and equipping of the water transport units. Another separate mountain water transport company model option is also possible. Without changing the TO&E, the its supply delivery capability can be increased by 55 metric tons if all water transport craft are switched to wire-framed inflatable rafts, getting rid of the catamarans (which carry 66.7% less than rafts). This model can used on high mountain rivers, as well as for class two rivers on which there are no difficult obstacles. Figure 4 shows the parameters that will determine a separate mountain water transport company's supply carrying capability in the second option.

On class one rivers, therefore, with the same number of personnel, the number of raft crews can be doubled by reducing them to two men, making it possible to increase the number of rafts to 58, with a total maximum carrying capacity of 435 metric tons, which is equivalent to two sets of such watercraft.

As is apparent from figures 3 and 4, the optimum composition of men, equipment, TOE, and their watercraft and carrying capacity are not a constant. The amount of military cargo that can be delivered depends on the class of the mountain river, watercraft specifications (number, type and volume of inflatable elements in their construction), and the raftsmen's technical and physical training.

The amount of cargo carried on mountain rivers may change even within the limits of a single operational area that encompasses mountain regions with varied physical and geographical conditions. For example, the rivers of East Pamir are level while those in West Pamir, which may run the same direction, are mostly whitewater and full of rapids. This affects the water transport units, whose thresholds (class 5) are laid out in the first Separate Water Transport Company (Mountain) model, figure 3.

The fact that the mountain forces and military units operate in varying directions and within the task forces' main advance makes another military formation transport model possible, Depending on transport volume, logistical factors and physical and geographical conditions, it could include: one-two water transport companies, one-two pack transport companies, a company of carriers, and combat, logistics and technical support units. Its three or five company (650 to 900 men) composition makes the military unit an independent tactical unit capable of delivering cargo and protecting, defending and safeguarding its assets without diverting combat formations from their primary tasks.

The separate mountain transport battalion organization and staffing structure is based on a modular structure in which a water transport company (any of the models or a combination thereof depending on logistical and physical and geographical factors) and a company of carriers are separate modules.

The entire separate mountain transport battalion or, depending on operational and logistical factors, separate detachments (modules) may deliver supplies, making it possible to maintain the transport capability of the other separate mountain transport battalion detachments (modules). Figure 5 shows an alternative way for separate mountain transport battalion/detachments to deliver supplies.

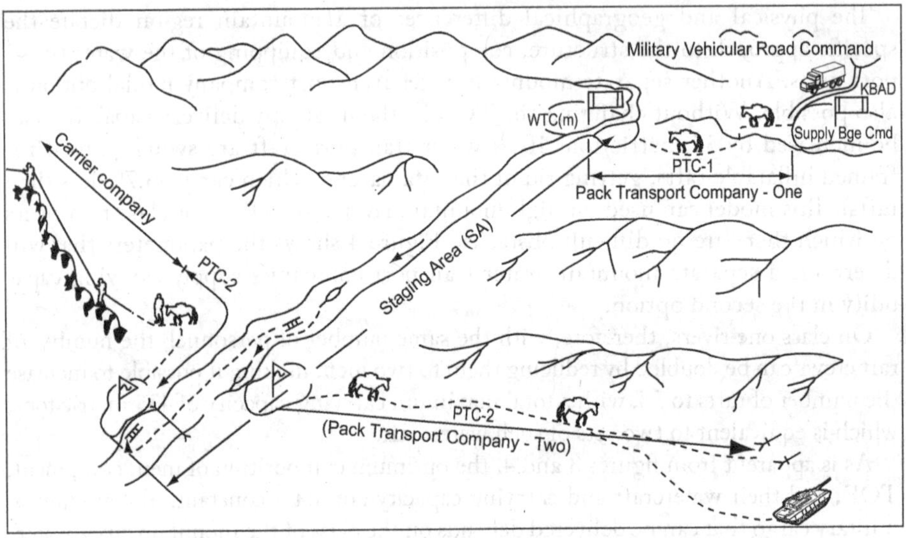

Figure 5 Separate Mountain Transport Battalion unit supply delivery methods.

The separate mountain transport battalion should include pack transport units to carry loads across passes that are inaccessible to vehicles. Their tasks will be: to deliver supplies from a higher rear echelon to the supply units of mountain forces and detachments or directly to their combat formations; and to transport supplies to final or halfway delivery points from the helicopter pads and airfield landing strips of military transport aircraft. That and to deliver supplies from a higher rear echelon to the staging area for further transport by water; deliver loads from the river (at the destinations) to where the brigade (regiment) or battalion is deployed; return the watercraft to the staging area after the trip; reinforce the water transport units to carry loads and watercraft by land when impassable rapids appear on the route during stage-by-stage transport; if the situation changes, maneuver the watercraft into lateral or adjacent gorges (new directions).

Unlike the separate pack transport company (SPTC) that existed in the Turkestan Military District prior to 1991, the structural model of the proposed separate mountain water transport battalion's pack transport company (PTC) requires supply platoons to switch to pack animal squads. This is primarily because of the nature of the mountain terrain and the limited capability of pack animal squads to move across rugged terrain. If necessary, a company's supply platoon can be reinforced, for example with a third squad of a third pack transport platoon. In this way, while maintaining the general parameters of the organizational staffing structure, a separate pack transport company becomes a military formation with much better capability thanks to internal structural changes.

Loads can be delivered to the river by pack from the field warehouses of formations (corps, army) located behind a pass, airfield landing strips and helicopter pads by placing the supplies on the ground or alternatively in watercraft.

Operational planning and execution need to correlate the capabilities of pack transport and water transport formations. For example, a pack transport company (PTC) that is part of an separate mountain transport battalion (213 pack horses that can deliver 17 to 20 metric tons of supplies depending on the elevation of the terrain and the lay of the land) roughly corresponds to the capabilities of a water transport company (WTC) on a class 5 river. This is 20 metric tons of useful load, not counting the weight of the crew, weapons and watercraft frames (figure 3). But on class one rivers, where a water transport company's capabilities increase sharply and can be 270-290 metric tons per trip (depending on the conveyance method—in piloted vessels or by raft), pack transport is less efficient than water transport by a multiple of 14 to 17, assuming parity of personnel.

This considerable disparity between a pack transport company and a water transport company capabilities in high mountain conditions reduces water conveyance volume to a minimum. In order for the units to be at par with each other, it will be necessary to use 3,600-3,870 pack animals whose maintenance at elevations above 3,500-4,000 meters, far from supply bases and with a constant need for fodder resupply, will cause additional difficulties and is, in our view, problematic.

At the same time, one should consider coordination between these military formations since pack transport can be used to the maximum in the preparatory period of an operation (battle) to stockpile supplies on the ground in the staging areas. By using a shuttle method and given a reserve of time, the number of service animals can be reduced to a single pack transport company. It will take six to 35 days to deliver the loads to the staging area at the specified volume (290 metric tons), with one-three trips per day, depending on the distance (5-30 kilometers).

An expert might question the use of pack animals and military rafting when there is modern technology such as helicopters. However, practice shows that it is difficult to use MI-8 MTV transport helicopters above 3,500 meters because of the thin air, which greatly reduces their cargo capacity. Their useful cargo capacity does not exceed 0.5-0.8 metric tons even at a minimum tactical radius of 50 kilometers and an ambient air temperature of up to 20°C. It is estimated that it will take 340 to 580 helicopter flights to deliver the load of the watercraft of a second model water transport company (58 rafts) when the river class is between 1 and 2. Even with drops (airborne delivery) or helicopters hovering at minimal height with no landing, it will take 290 helicopter flights. A comparison of these statistics casts doubt on these helicopters' ability to collaborate effectively with water transport formations on high mountain plains.

In order to resolve the challenges of delivering supplies to high mountain plateaus, military transport aircraft can be used in the assembly areas of water transport military units and detachments at the staging areas. These aircraft could deliver supplies from frontline or center depots, thereby avoiding intermediary logistical channels—army and corps logistics—and possibly frontline logistics as well.

Military transport aircraft can deliver supplies to high mountain plains by landing (without turning off the aircraft engines) on unpaved air strips or by parachute. This experience was gained back in Soviet times, when East Pamir had the

USSR's highest altitude airfield in Murgab with an elevation of over 3,600 meters. Transport aircraft flights have currently been resumed to hard-to-reach areas of the Pamir, specifically Murgab, in AN-26 aircraft, whose flying specifications make it possible to deliver up to a five-metric-ton load to such elevations over a distance of more than 700 kilometers.

The use of water transport formations in mountain and coastal areas will enable all logistical elements to meet another serious challenge – ferrying supplies, vehicles and pack transport across the numerous mountain rivers. After all, in a combat situation a retreating opponent or his special operations groups operating in the rear of our formations and military units are going to blow up bridges, which is why water transport formations make the military units and logistical elements more mobile since the watercraft used on mountain rivers make it possible to set up pontoon bridges, and assault and ferry crossings.

Let us look as an example at the exercises in which the Republic of Uzbekistan's Ministry of Defense special operations units participated. The units demonstrated how to set up a pontoon bridge from catamarans that took part in operational rafting during the exercises. It took 45 minutes to assemble, set up and secure a 15-meter long bridge to move the pack transport across a mountain river gorge.

The proposed water transport formations (mountain) can also be used to maneuver pack transport units on class one mountain rivers by placing the animals on wire-framed inflatable rafts (WIR) with appropriate area (deck) setup. This can greatly reduce the energy used by the pack animals as compared what they would use if they had to cover those distances over the mountains on foot. There have been instances of equipment and large domestic animals being transported considerable distances—200-250 kilometers—on rafts along mountain taiga [subarctic forest] rivers. The deck area of a 3x7.5 meters WIR raft can hold up to six pack horses, that is, as much as a KamAZ-4310 truck.

The difficult mountain terrain and numerous rock projections and rock falls make it hard or impossible to deliver supplies by pack transport. This is why a separate mountain transport battalion needs a carrier company that can carry 3-3.5 metric tons of supplies. Its purpose is not just to take loads from the river up to where the supply companies of the mountain forces and military units are deployed or at times to the supply platoons and firing positions of mountain battalions. This separate mountain transport battalion has much broader functions, first and foremost to reinforce water transport units when supplies are being prepared for military rafting—to waterproof the loads and take them from the packing site to the river in the staging area.

It also has to provide stage-by-stage and area-by-area transport that could require additional forces to carry supplies around impassable rapids, and to do the same with watercraft during stage-by-stage transport in order to expedite delivery.

It also has to return the watercraft to the staging area when there is no other way to transport them such as during area-by-area transport when one set of watercraft is being returned to the loading point while another is transporting supplies further to the offloading sites.

It also has to perform search and rescue for loads and watercraft that were sent unmanned and did not reach their destination at the expected time (grounded, stuck on an obstacle, caught in a vortex, etc.). It also carries out the orders of the senior officer as the logistical situation unfolds.

A separate mountain water transport company and a separate mountain transport battalion should have regular vehicle units in order to make them more mobile and autonomous. For example, a company has a vehicular platoon made up of a personnel transportation squad, a transportation squad (mixed) for food, materiel, petroleum, oil and lubricants (POL), and a maintenance and repair squad—a total of 17 to 20 vehicles. And a separate mountain transport battalion includes a vehicle company made up of personnel and service animals transportation platoons, as well as two-three squads to transport food, materiel (one-two squads) and POL. The number of vehicles could vary widely depending on the number of major units. For example, the fleet could number 95 to 100 vehicles for a regular three company strength (water transport company, pack transport company, and motor vehicle transport company).

Given the nature of modern military operations, it is a priority to make logistics support and its structural elements more robust and viable. This also fully applies to the proposed water transport and mountain transport detachments and military units. It is important to equip them with armored personnel carriers, portable anti-tank rocket launchers and anti-aircraft rocket launchers, communication and reconnaissance assets, and engineer explosives. It follows from this that a separate mountain transport battalion should have a company to protect, defend and safeguard the battalion, which would include the combat support platoons.

Thus, despite its seemingly unwieldy organizational structure, a separate mountain transport battalion is a universal transport formation that can take supplies to the most remote and hard-to-reach mountain regions. It is also universal because the units are capable of maneuvering independently and protecting and defending its forces and assets, often without having to be reinforced by other logistics and combined arms units and detachments. With their own vehicles and tanker trucks, water and pack transport companies and motor vehicle units can cover large distances of 1,000-1,500 kilometers in the mountains with two-three fuel loads (in fuel tanks, the tanker trucks and additional containers in the cargo area). After the work is completed, the same vehicles can be used to transport supplies from the depots of task forces, and sometimes combined forces, to water transportation companies and pack transportation companies for further transport. If necessary, pack transport can redeploy water transport units into cut-off rafting areas while livestock rafts enable pack transport units to maneuver around difficult mountain passes.

Let us cite for comparison purposes the size of the tactical military formations of foreign armies as we consider how to optimize the composition of the proposed separate mountain transport battalion. In Germany's ground forces, mountain-infantry, transport, supply battalions number up to 900 men. Medical transport battilions have more than 1,000, while pontoon battalions have roughly 800 personnel.

The above demonstrates that corps (army) logistics needs to have two-three regular separate mountain water transport company or separate mountain transport battalions in the areas of operation of first echelon forces and military units of the army corps (army) that are operating in mountain districts and regions with a developed hydro-graphic network that allows use of water transport units. Depending on the length and classification of the mountain rivers and the delivery amount and deadline, these forces and assets could be sent to reinforce the rear of a first echelon motorized rifle division or motorized rifle brigade (mountain) in order to make them autonomous (*Armed Forces of the Main Capitalist Countries*, M., Voyenizdat, 1988, pp. 163, 172).

France's alpine ground forces infantry battalion has roughly 900 men and up to 130 vehicles (ibid., pp. 213, 217). Italy's separate alpine ground forces brigade numbers 5,777 men, and an alpine battalion, 950 (*Ground Forces of the Capitalist Countries*, M., Voyenizdat, 1980, p. 400).

Given the tasks of the military units and detachments (deliver supplies; maneuver supplies, forces and assets; evacuate the sick and wounded, damaged weapons and materiel; and arrange mountain river crossings), separate mountain water transport company and separate mountain transport battalion functions should be regarded as a key task of logistical rather than material support. The transportation departments of the task forces' logistics administrations (corps, army, front) need to set up mountain transport offices to coordinate the actions of these specialized transport formations in order to strengthen the logistics groups and increase their capability to provide timely and comprehensive support for combat in hard to reach and isolated theaters of operation.

The deputy commanders in chief (commanding officers) of logistics task forces must have a reserve of mountain water transport formations to make rapid supply deliveries to troops in theaters of operation with a developed hydrographic network.

Given the large spatial scope and depth of present-day operations in mountain regions, there may be a need for mountain transport brigades (MTB) that could include the proposed separate mountain transport battalion forces and assets for operational, logistical and technical support, as well as a helicopter assault squadron with airfield maintenance units. Based on the modular principle, being highly mobile and with a high degree of tactical independence, these specialized mountain transport forces could consistently deliver supplies in isolated theaters of operation, reacting flexibly to a rapidly changing situation. Final conclusions as to the advisability of the proposals made in this article can be drawn following special military research and operational command and staff exercises in a mountain area.

COMMUNICATIONS

20

Mountain Combat: Hard to Move, Hard to Shoot—Even Harder to Communicate

---

LTC Lester W. Grau, US Army (ret.) and Lieutenant Jason Falivene USMC

Originally published in *The Journal of Slavic Military Studies*, September 2006.

Communicating in mountains is a challenge since there are few ideal spots for communication. FM radios, which are line of sight systems, frequently cannot communicate because their signals are absorbed by terrain folds and features. If all the force is on the same side of the mountain and the mountain forms a bowl, FM communications are usually possible. However, radios located on the same side of the mountain at different altitudes have difficulty communicating because of intervening terrain and communications dead space. If the force is deployed on the same side of a mountain which curves out, communications are especially difficult. Even FM radios located on the summit of the mountain have difficulty communicating with radios located further down the mountain slope due to dead space. Communications sites must be carefully selected—and often become key terrain. When lineof-sight communications in mountains are possible, communications are excellent, but there are few sites where line-of sight is possible to all other elements in the net. There are often only three solutions–either move the radio to where it can communicate, set up a radio retransmission site, or relay messages across the net.

Radio retransmission sites are expensive in terms of personnel and equipment. TO&Es normally do not provide adequate personnel and equipment to provide several retransmission sites. Further, since the retransmission team must work away from the main body, it must have enough personnel to protect itself and haul all its gear to the retransmission location. Batteries, antennas, guy wires, rations, water, weapons, ammunition, and personnel gear are heavy. Moving a site is labor intense. Maintaining a site is also a chore. Fresh batteries, chow, and water have to be carried to the site and personnel rotated. If the mission is not static defense, the retransmission site has to constantly shift to yet another site where it can adequately support its

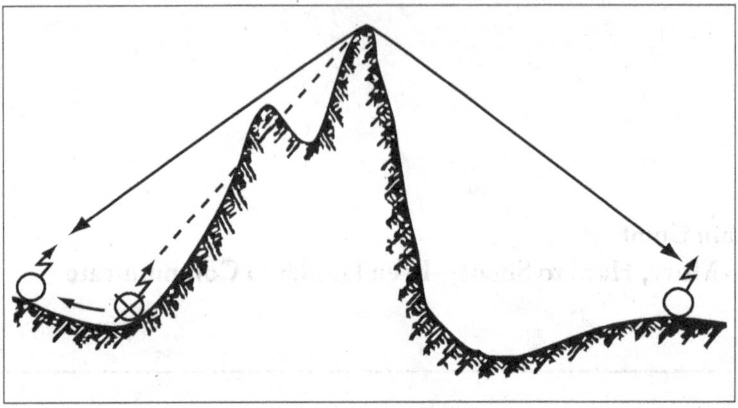

Figure 1 Deploying radios to direct the transmission over a narrow mountain peak (both radio operators must be able to see the mountain peak).

unit. Such sites are not easy to find—or reach—yet the communications teams must keep up with the advancing force which is usually moving along easier terrain.

The Soviets in the mountains of Afghanistan

During the Soviet-Afghan War, Soviet forces often entered the massive Hindu Kush mountain chain or the imposing Sulieman range. Radio retransmission sites were essential. The Soviets often used Mi-9 VZPU command and control helicopters or other helicopters to conduct retransmission support during movement.[1] Often, the Soviets lacked sufficient personnel and equipment to establish enough ground-based retransmission stations. Then, the Soviets had to resort to radio relay–a long, tedious process involving relaying the message to various stations until it eventually reaches the intended recipients. At first, communications troops made errors during radio relays. The Soviets solved this problem by requiring their communicators to physically record all messages prior to relaying them. Then they trained their communicators to write clearly and quickly on standard forms using capital letters while never lifting the pencil from the paper. The communicators would repeatedly listen to various transmissions and record them to gain proficiency. Over time, these simple drills improved their accuracy and relay time significantly.[2]

UHF radios also present problems in the mountains. Like FM, UHF signals are absorbed by intervening terrain, yet UHF are not restricted to line-of-sight and can bend somewhat over mountain tops. The Soviet tactical UHF radios were normally able to communicate out to 100 kilometers over open ground. They could also communicate out to 100 kilometers with an intervening mountain as long as the transmitting and receiving stations were on high ground and the intervening mountain was midway and no higher than 200 meters above the stations. Taller mountains and multiple peaks interfered with UHF communications. A single, closer, yet lower,

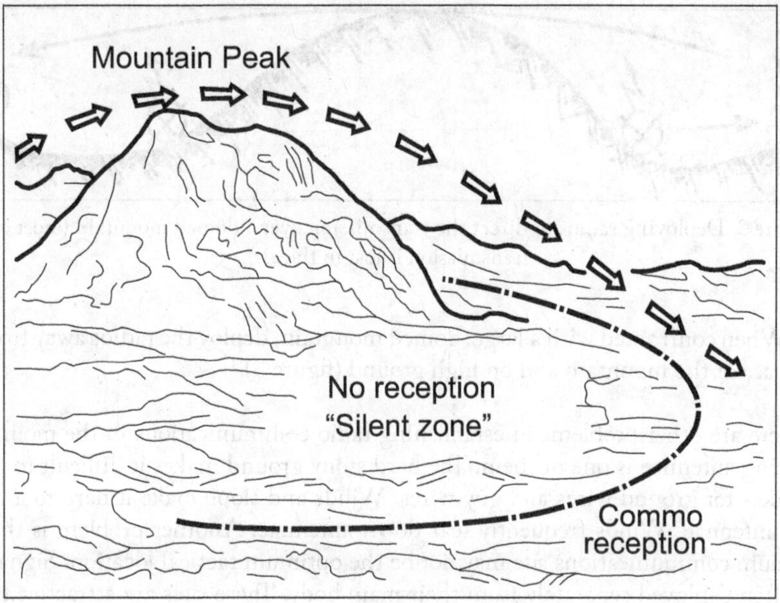

Figure 2.

peak cut transmissions to 20-22 kilometers and that was only if the mountain crest was narrow and both stations were aimed at the sharp peak. UHF communications distance was cut to 10 12 kilometers if the intervening peak rose up to 100 meters higher than the stations. If there were a series of 200 to 400 meter peaks between stations, transmission distance was cut to 9-10 kilometers–and only if both stations were far enough away from the mountain bases and used whip antennae. Large, domed mountains cut UHF transmissions to 5-6 kilometers, while cut-up rugged mountain terrain further limited transmissions to 4-5 kilometers. UHF communications were frequently lost while moving along mountain roads or in the "silent zone" on the far side of mountains.[3]

The Soviets took various measures to support UHF communications in the mountains. These measures included:

1. Select communications sites that have a narrow single mountain crest between them. Aim the transmissions at the highest peak. Keep the sites away from the mountain base (figure 1).
2. Deploy radios away from the mountain base to a distance at least equal to the distance of the slope between the base and mountain crest. (figure 2).
3. Deploy radios to commanding heights to improve their line-ofsight to the top of the intervening mountain.
4. Deploy the radios where they can communicate over a single mountain rather than a series of peaks and defiles.

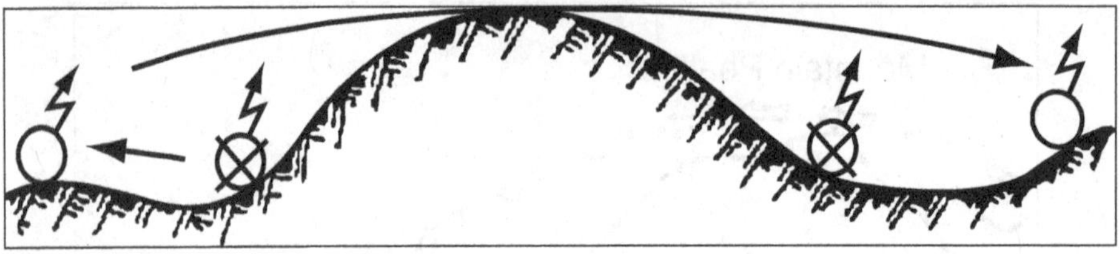

Figure 3 Deploying radios to direct the transmission over a domed mountain (most of the transmission is lost in the air).

5. When confronted with a large, domed mountain, deploy the radios away from the base of the mountain and on high ground (figure 3).[4]

There are other problems in establishing radio communications in the mountains. Erecting antennae is one of them. The hard stony ground makes it difficult to pound in stakes for ground wires and guy wires. Winds and slope make it hard to aim and tune antennae. Winds frequently tear down antennae. Another problem is that the optimum communications site may not be the optimum tactical location. Signal sites are often deployed separately from their main body. These sites are attractive targets that do not have a lot of combat power. Weather is another problem. Mountains get more than their fair share of thunderstorms, ice storms, and snow. Antennae attract lightning. Antennae ice over rapidly and the ice decreases the transmission power significantly. Ice has to be removed, but this is not easy to do on a long antenna on a mountain during a snowstorm. Diesel engines do not run very well at high altitude, yet most communications generators are diesel-fueled. Standard radio batteries do not handle cold well and therefore the more-expensive lithium batteries are necessary in the high mountains. Finally, communications personnel need to be relieved, rotated and rested regularly in order to maintain good communications.

The US Army in the mountains of Afghanistan

The US Army experienced the difficulties of communicating in the mountains of Afghanistan. Although the it has far more satellite communications radios (SATCOM) than the Soviets did during the Soviet-Afghan War, the Army experienced many of the same difficulties. During operation Anaconda, 101st Air Assault and 10th Mountain Division units formed a single brigade—Task Force Rakkasans. TF Rakkasans found communications a challenge, but they were able to make them work.[5]

The base radio for the ground forces was the SINCGARS family of FM radios. Since the force landed inside a mountain bowl and the range of the battlefield was not too great, the FM radios worked surprisingly well, however, terrain folds frequently absorbed signals. "If you can't talk, move" was the working solution, although some wags observed that "communications drives maneuver." The fire direction net, brigade command net and battalion internal nets were all on FM radio. The task force did not use the frequency hopping option on the SINGARS since they also talked to neighboring special operations forces (SOF) on FM on a single frequency. During the two-week operation, the task force changed its frequency only once — and this was due to constant interference and bleed over from another net. A major advantage of FM radio was that ground forces could communicate with helicopter aviation once they were flying in the bowl. However, once the helicopter cleared the crest of the bowl, FM communication was lost.

The helicopters talked to each other on UHF radio. The ground forces had little luck with UHF radio. Unlike SINCGARS, UHF traffic was plain text. The pilots could talk to the main headquarters at Bagram (over 100 miles away) on UHF. TF Rakkasans had little confidence or success using UHF on the battlefield, nor did they use HF radio. Canadian forces and SOF used HF radio to send scheduled reports, but not for combat. The ground forces did not bother taking VHF radios since they considered VHF as "unreliable and too complicated", and "big and bulky and useless." Indeed the 60-pound weight of the issue VHF radio is prohibitive on any terrain.

The AN/PSC-5 TACSAT radio was the primary means of communications beyond the mountain bowl during Operation Anaconda. Encrypted satellite communications were reliable, but the narrow-band width assigned to ground forces by the DAMA (demand assigned multiple access) system made communications very slow and hard to understand. Three battalions and the brigade had to share one 25 kilohertz channel! Further, the brigade's TACSATs were not data capable which frustrated speed of communications and accuracy. The USAF and SOF, on the other hand had broad-band TACSAT and enjoyed good communications. If no helicopters were in the bowl, TF Rakkasans had to contact the AWACs aircraft by TACSAT. Since AWACS lacks TACSAT retransmission capability to helicopters, AWACs would manually relay messages to the helicopters.

Other means of communication were Iridium satellite telephones. Although they are difficult to encrypt, they provided excellent emergency communications and allowed the brigade to enter SIPERNET through laptop computers. Much necessary communication was done on the Internet through the SIPERNET-Iridium connection. Further, the Iridium net transmitted and received at normal speed while the TACSAT net was very slow and hard to understand. No wire communications were used, since wire is heavy and the brigade had limited lift capability. Radio batteries lasted about a day and fresh batteries were a key logistics concern.

Task Force Rakkasans had two TACSAT narrow-band nets (one each from the 101st and 10th divisions), a USAF broad-band TACSAT net used by the Air Liaison Officer (ALO) to talk to supporting high-performance aircraft, an FM fire direction net and a FM command net. There were no brigade administrative and logistics or intelligence nets due to the limited number of TACSAT nets available. In order to save time and insure accuracy, when the brigade commander spoke, he spoke only to his commanders and everyone else stayed off the net. Due to the heavy fighting, there was no command and control helicopter over-flying the battlefield. The brigade staff worked out of Bagram where they had access to the Predator UAV feed coming into the 10th Mountain Division Headquarters. Each battalion had two TACSAT radios and the normal compliment of FM radios. Battalions were on the brigade TACSAT command net, the brigade FM command net, the brigade fire direction net and internal command net.

Special Operations Forces had quality FM, UHF and wide-band TACSAT radios which provided good communications throughout the operation. Air Liaison Officers had good communications with their wide-band TACSAT radios throughout the operation.

Operation Anaconda highlighted the problems with DAMA and demonstrated the need to issue broad-band TACSAT radios to conventional forces. It further demonstrated the need for data-capable TACSAT radios and more satellite coverage. It also showed the need for TACSAT radios in helicopters.

Roger. Out

Despite dramatic advances in communications technology, communicating in mountains remains a problem. Satellite communications systems provide greater capability, but they have problems operating around terrain folds as well. Bandwidth and lack of data capability are further serious drawbacks. There is a role for FM and UHF. Iridium phones with computer data link are particularly valuable. Much of the staff work and battle management was accomplished in secure chat rooms. The problem with chat rooms, however, is that anyone with ac- cess can join in. The siren call to participate in an operation, even remotely, brought a lot of strap-hangers and time-wasters into the chat room.

Operation Anaconda demonstrated the need to have over-thehorizon communications with aircraft and the main headquarters. Further, the operation demonstrated the need for a survivable command and control aircraft. Data burst technology was not available.

Communications in the mountains is possible but requires planning, training and experience. It also requires better equipment and improved bandwidth.

Notes

1. *The Russian General Staff* (translated and edited by Lester W. Grau and Michael A. Gress), *The Soviet-Afghan War: How a Superpower Fought and Lost*, Kansas University Press, 2001, 210-221.
2. Yu. Kuzimichev and S. Ponomarev, "Perepriem i retranslyatsiya v gorakh" ("Relay and Retransmission in Mountains"), *Voyenniyy vestnik* (*Military Journal*), December 1986, 74.
3. "Radiosvyaz' na YKB-stantsiyakh v gorakh" ("UHF Radio Communications in Mountains"), *Armeyskiy sbornik* (*Army Digest*), February 1997, 44-45.
4. Ibid, 45.
5. This section based on an interview with Captain James Riley, Signal Officer, 3d Brigade, 101st Air Assault Division at Kandahar, Afghanistan on May 13, 2002 and an interview with Captain Francisco Ranero, Signal Officer of 1-87th Infantry Battalion, 10th Mountain Division at Fort Drum, New York on June 11, 2002.

MEDICAL CONSIDERATIONS

21

Medical Implications of High Altitude Combat

LTC Lester W. Grau (ret.) and William A. Jorgensen, DO MC USAR

Originally published in the *US Army Medical Department Journal,* April-June 2003.

The fact that a piece of land is inaccessible, uninhabitable or of little practical value is no guarantee that nations will not fight over it. Long, bloody wars have been fought and are being fought for inhospitable mountain real estate located between 10,000 and 23,000 feet [3,050 and 7,015 meters] in elevation. Examples of such high altitude combat include the 1953-1974 Chinese invasion of Tibet and subsequent guerrilla war, the 1953-1958 Mau-Mau rebellion where British troops fought rebels in the Aberdares mountains of Kenya, the 1962 Sino-Indian War in the Himalayan mountains bordering Bhutan and Tibet, SovietMujahideen combat in Afghanistan's Hindu Kush mountains from 1979 to 1989, the Peruvian government's clashes with Sendero Luminoso guerrillas in the Andes throughout the 1980s and the Indo-Pakistan continuing conflict over the ownership of the Siachen glacier which began in April 1984. Recent (1999) Indo-Pakistani clashes in the Kargil area of disputed Kashmir again demonstrate that high altitude combat is often contemporary combat. Tens of thousands of combatants have perished in inhospitable ice, snow and rock while battling for national prestige, water rights, survival or geographic positioning.

The United States Army has not had to fight at such altitudes, but with the war on terror, the possibility of United States military commitment to such areas is not all that remote. Operation Anaconda in Afghanistan is the highest altitude ground fight [10,500 feet] in US history. Since the Army is still inexperienced fighting at these altitudes, it should draw from the experience of others. There are some distinct medical problems that medical personnel should plan for in the event of a high-altitude contingency operation.

High-Altitude Medical Considerations

The world's highest mountains are in the Himalayan and Karakoram mountain chains of Asia. The Himalayan Mount Everest towers at 29,028 feet [8,853.5 meters] whereas the highest point in the United States, Mount McKinley in Alaska, is 20,320 feet [6,197.6 meters]. The highest point in the Colorado Rockies is Mount Elbert at 14,433 feet [4,402.1 meters].The highest point in the European Alps is Mont Blanc at 15,771 feet [4,810.2 meters].[1] Man is not naturally designed to live and work at these high altitudes. Any time a person travels to an altitude of 8,000-10,000 feet [2440-3050 meters] or higher, the atmospheric changes in pressure and available oxygen cause physiological changes which attempt to insure that the body gets enough oxygen.[2] These physiological changes are pronounced among mountain peoples, who have lived in the cold and higher altitudes for generations. Their bodies are short, squat, stocky and barrel-chested when compared to those of lowlanders, Their hands and feet are stubby. Their hearts are bigger and their bodies contain 20-percent more red blood cells. Their red blood cells are larger than those of lowlanders. Their heartbeat is slower and their capillaries are wider. The alveoli in their lungs are more open for oxygen absorption. Many develop a fatty epithelial pouch around the eyes to counteract cataract and snow blindness.[3]

High altitudes are characterized by extreme cold, strong winds, "thin" air, intense solar and ultra-violet radiation, and rapidly changing weather including severe storms which can cut off contact for a week or longer. Personnel should be screened and acclimatized before deploying to high altitudes since these conditions at high altitude are usually more dangerous than enemy fire. Medical personnel should prepare for the special demands of high altitude treatment and care. Bullet and shrapnel wounds, which normally are not serious, can quickly prove fatal at altitude. Movement in the high mountains often results in broken bones, severe lacerations and contusions and internal injuries caused by falls and falling rock. Frostbite and hypothermia are a constant danger. Acute mountain sickness, high altitude pulmonary edema, and cerebral edema are frequently fatal consequences of working at high altitude. Mental and physical abilities decrease at high altitude and high altitude also induces personality disorders. Sudden weight loss is often a problem. The rarefied atmosphere permits increased ultra violet ray exposure which create problems with sunburn and snow blindness. High-altitude shelter heating is often by unvented kerosene stoves, which means that personnel breathe air which is thick with soot.

Medical personnel will be exposed to the same dangers of working in high altitudes and much of their normal medical equipment will not function, or function effectively, at high altitudes. For example, hospital generators and vehicles are often diesel-powered. Diesel engines lose efficiency at 10,000 feet [3,050 meters] and eventually do not work at all due to the thinness of oxygen at higher elevations. Helicopters cannot haul heavy loads over 13,000 feet [3,965 meters] as their rotors lack thick enough air to "bite" into. Altitude requires additional animal or gasoline-fueled overland transport which adds to the physical demands and logistic requirements of medical support in this environment.

Screening and Acclimatization

At high altitude, there is less oxygen and atmospheric pressure. The soldiers selected for high-altitude duty should be screened for their ability to function in this environment. Soldiers should be in excellent physical condition and have sound hearts and lungs. Short, wiry soldiers are preferable to tall, over-muscled soldiers. Selected soldiers should possess a higher level of intelligence in order to allow them to more-readily adapt to the trying terrain.[4] Personnel who have had radial keratotomy (RK) corrective eye surgery should not go "to altitude" as their vision may permanently cloud. Personnel records should be screened for previous high-altitude sickness. Some personnel can be administered acetazolamide (diamox) prophylaxis, however, personnel with sulfa allergy or G6PD deficiency cannot use acetazolamide. Personnel with the sickle cell trait should be excluded since rapid exposure and dehydration could set them up for splenic syndrome. Further, certain medications (i.e. any benzodiazepine such as valium) inhibit acclimatization and personnel using these should be carefully evaluated.[5]

All personnel should undergo an acclimatization program to accustom them to their new environment and to improve their respiratory and cardio-vascular systems. A physically-fit soldier can adapt to the cold in about three weeks.[6] Experience shows that the body normally adapts to a new altitude in about two weeks time. During the acclimatization phase, the body accumulates additional red blood cells which help transport needed oxygen.[7] The Pakistani Army acclimatizes their personnel over a seven-week cycle. They begin with a three week stay at 10,000 feet [3,050 meters], where personnel acclimate to the cold while they undergo daily physical conditioning, and learn mountaineering, rock climbing, rope rappelling and mountain survival. During the final four weeks, the soldiers learn advanced mountaineering techniques, trek to 14,000 feet [4,270 meters], return and trek to 17,000 feet [5185 meters] and then return and finally trek to 19,135 feet [5,836 meters].[8] Despite all training and efforts, acclimatization is not possible at heights over 18,000 feet [5,418 meters], so personnel exposure at these heights must be limited and closely supervised.[9]

Medical personnel should advise logisticians and planners on special considerations for high altitude combat. For example, lightweight, pre-cooked, high-caloric rations that are high on carbohydrates are essential and aid acclimatization. Supplementary candy and soups will help off-set the inevitable loss of appetite at altitude. Boiling water from snow for purification takes an unwarranted amount of fuel and so provisions have to be made to provide water or water purification equipment to the soldiers "at altitude" where dehydration is a constant threat. NCOs need to "push water" to compensate for the diuretic effects of acclimatization. Troops working above 15,000 feet should be issued pressurized sleeping bags. These bags, which are inflated with a foot pump, have been tested to provide equivalent pressure of 8,370 feet [2,550 meters] while at 13,600 feet [4,150 meters].[10] Medical personnel should further recommend a rotation program where highest-altitude exposure is limited to ten to fourteen day increments.

During the Soviet-Afghan War, Soviet doctors and physicians' assistants often accompanied small units on high-altitude missions, since the patients required immediate medical care and evacuation (medevac) took too long to save many patients.[11] Limited medical staffing in the US Army will prevent many doctors and physicians assistants from accompanying the high-altitude patrols. Therefore, the brunt of the responsibility for saving injured and sick soldiers will fall on the combat medic. In addition to the medic's normal skills, he will need to be trained in mountain rescue techniques, treatment of altitude-specific medical problems and high-altitude medevac procedures. The medic should accompany the unit during the acclimatization process and rotate in and out of the high-altitude area with the unit.

In an emergency, diamox can be given to nonacclimated personnel (125mg twice a day) starting the day before ascent and up to two days after ascent. However, this is an emergency measure that should only be used for a forced ascent to over 10,000 feet in one day. Normal acclimatization is preferable. There are side effects to diamox, such as peripheral paresthesias (numbness in the extremities) and bone marrow suppression. The possibility of bone marrow suppression is relatively rare, but cannot be ignored.[12]

Frostbite

Frostbite is the most common injury at altitude. Frostbite is a continual danger, but especially so following any exertion. Sweat rapidly freezes around the toes and fingers. Frost bite may be classified as frostnip, superficial frostbite, or deep frostbite, depending on the severity of the case. Frostnip usually occurs on the tips of the ears, nose, fingers, toes and cheeks and it noticeable as a whitening of the skin. Simple warming of the area is usually sufficient treatment. If it advances to superficial frostbite, the affected areas will be firm and have a white waxy appearance. Warming and gentle massaging of the area are the necessary treatment. As the area rewarms, it may turn a mottled blue or purple and swell. Nerve damage may also accompany superficial frostbite. In case of deep frostbite, major areas of tissue are frozen and killed. The areas are cold, pale solid and hard. Infection and amputation often result. The patient must be evacuated.

Medics should be cautioned that once the frozen area is thawed, do not allow it to refreeze and do not to thaw unless continual warmth and litter evacuation are available. It may be necessary to prevent thawing in order for the injured soldier to walk out. Once thawing occurs, the severe pain prevents the patient from walking out, although codeine, aspirin or morphine should help the patient.[13] Evacuation at altitude is often difficult. Weather or weight limitations may prevent a helicopter from flying to the patient. Often, patients must be carried on stretchers to lower elevations where the helicopters can fly to. Soviet experience fighting in the mountains of Afghanistan proved that 13 to 15 men might be involved in carrying out one patient. Exertion at altitude is difficult and the stretcher party had to provide its own security as well.[14]

Hypothermia

Hypothermia is the result of the body losing heat faster than it can produce it. The body's core temperature begins to drop and the patient shivers violently, has trouble using his hands and is generally clumsy. When the core temperature falls to 90o-95o Fahrenheit [35o-32o Centigrade], the patient becomes uncoordinated, has difficulty speaking and is disoriented and apathetic. As the core temperature continue to lower, the patient becomes more irrational, lapses into semiconsciousness and eventually unconsciousness and cardiac arrest. If the patient cannot be re-warmed on site, the patient needs to be evacuated. Medics should be equipped with the mountaineering "hydraulic sarong," a re-warming device that wraps around the patient and circulates a warmed liquid from a camp stove or catalytic generator around the patient's body. "Hot oxygen" breathing units, which use a soda lime and CO^2 reaction to warm oxygen, can also aid in re-warming the body core.[15] When a hypothermia casualty's body core temperature drops below 90o, when he stops shivering or when he passes out, extra care must be given to handling him or he may develop cardiac arrhythmia and sudden death.[16]

Falls and Climbing Injuries

Fractures, severe lacerations and contusions or internal injuries often result from falls or by being hit by falling rock. The ABCD's of trauma should be followed, and the patient examined for spinal injury as one of the first checks. Medics should not hesitate to put a cervical collar on fall victims with suspected cervical spine injuries, particularly since these can usually be cleared in the field, avoiding an unnecessary/ hazardous medevac.[17] Field splinting and immobilization should also be done before the patient is moved. The spleen, liver and kidneys are the most likely organs to rupture and bleed internally from a fall. A torn diaphragm or intestine is also a possibility in falls and climbing injuries.[18]

Mountain Sickness, High Altitude Pulmonary Edema, and Cerebral Edema

Mountain sickness or altitude sickness normally begins as a headache and may include insomnia, loss of appetite, vomiting, cough, shortness of breath, irregular breathing, tightness in the chest, loss of coordination, swelling around the eyes and face, general weakness and reduction in the volume of urine produced. The patient will lose physical coordination and mental acuity and tire quickly after mild activity. Mountain sickness normally takes at least 24 hours to develop but non-acclimated personnel often develop the symptoms within 6 to 12 hours (if they are quickly transported to elevations at 11,475-14,750 feet [3,500 4,500 meters]). Treatment involves awareness of potential problems, rest, sleep and a good meal. Should that fail, the patient should descend to a lower altitude for a few days rest until he improves.[19]

Moderate mountain sickness involves the same symptoms, but their intensity increases and urine output is often more reduced. If a day of rest does not help the patient, he should be brought down immediately. Usually, an early descent means an early recovery.[20]

Severe mountain sickness occurs in about 2-3-percent of mountain sickness cases and involves high altitude pulmonary edema and/or cerebral edema. Twenty percent of acute severe mountain sickness cases prove fatal. Signs and symptoms of high altitude pulmonary edema, the collection of fluid in the lungs, include persistent cough, gurgling chest sounds, red frothy sputum, breathlessness, and tachypnea and tachycardia. Younger soldiers (under age 25) and soldiers with a history of pneumonia or other respiratory illness are prone to high altitude pulmonary edema. The symptoms of cerebral edema, swelling of the brain due to fluid, include headache, difficulty in balance, loss of coordination and labored breathing. Severe mountain sickness may prove fatal within a few hours. Proficient medics may administer nifedipine and dexamethsone and, if available, administer oxygen or use a Gammow bag (pressurized bag also known as a certic bag).[21] The patient still needs to be brought down. Oxygen, Diamox, Tylenol, aspirin, codeine, Decadron, Valium, Lasix, Phenergan, or morphine have all been used to help the patient during descent.[22]

The best prevention of mountain sickness is a gradual ascent with plenty of fluids and food provided to the soldiers. Climbing soldiers need to avoid overexertion. The worst approach is to drive or fly the soldiers from low to high altitude and then let them out to finish the ascent.[23]

Combined Physician's Curriculum for Mountain Rescue and Combat[28]

Topic	Lecture	Practical Exercise
Introduction to goals, missions and content of high altitude medicine.	2	2
Special features of high-altitude physiology. Acclimatization to high altitude: Short term and long term. Mountain pathology.		
Acute mountain sickness: etiology, pathogenesis, clinical picture, treatment and prevention. Pulmonary edema.	2	2
Effects of high altitude on the nervous system, heart vessels, excretory system, gastro-intestinal tract, and circulatory system. Diagnosis, treatment and prevention.	2	2
Peculiarities of the course of "normal" illnesses at high altitude	2	2
Directing an acclimatization program. Personnel selection and prognosis of their health in the mountains. Medical oversight and preventive medical examinations.	2	2
Examination	—	2
Special features of mountain trauma. First aid for accidents	—	2

Topic	Lecture	Practical Exercise
Diagnosis of trauma, broken bones of the extremities, methods of moving immobile patients	2	4
Evacuating patients in the mountains	—	4
Wounds, methods of stopping hemorrhage, bandaging	2	6
Depression	4	4
Traumatic shock, first aid for shock from mountain accidents and trauma	2	4
Emergency medical service and transport of patients with trauma to the head, spine, chest, stomach and pelvis	2	6
Special features of medical treatment of freezing, frost bite and snow blindness	2	
Emergency medical service for drowning victims	2	
Examination	—	2
Special aspects of organizing medical support for forces in the mountains	2	
Rigging medical gear for evacuation and treatment in the mountains	2	2
Screening soldiers for service in the mountains	2	4
Oxygen equipment and its use	2	4
Training drills for high-altitude treatment	—	20
TOTAL	34	74

"Siachen Syndrome"

The change in barometric pressure and reduced quantity of oxygen at high altitude leads to mental status changes as well as physiological and psychosomatic changes. The Pakistani Army has noted that for every rise in a thousand feet, a person's temperament may change. A good-natured soldier at 19,000 feet may become irrational and selfish at 20,000 feet, introverted at 21,000 feet and unhinged at 22,000.[24]

Although not recognized as a disease, the so-called Siachen syndrome has been noted among veterans of fighting on the Siachen glacier. Its symptoms include disorientation and different psychological disorders. The experience has resulted in psychiatric treatment for some of the veterans.[25] Team-building, discipline and productive activity help prevent the apathy which leads to Siachen syndrome.[26]

Training the Medical Force for High-Altitude

The Soviet Union had a special course to train doctors to function effectively at high altitude. The course was founded in 1987, the seventh year of the Soviet-Afghan War, and was taught at the Kirghizistan medical institute alongside normal medical courses and prepared military and civilian doctors for mountain rescue and high-altitude treatment duties. (see figure). The course met twice monthly in hour-and-a-half sessions. The course devoted 34 lecture hours and 74 hours of practical application to medical topics. Another 792 hours were devoted to mountaineering training, of which 47 were lecture and the remainder practical application.[27] Although it would be difficult to find the time to train US military medical doctors to the same standard, the medical topics taught at the course may prove of value when planning a training course for medical personnel who may serve at high altitudes.

Medic's training for high altitude combat will necessarily involve many of the same skills needed in mountain search and rescue units. Medics will have to know how to rig patients and litters for evacuation from precarious positions. The search and rescue community has a wealth of information that can be tapped for military medical use. Search and rescue personnel in the Yosemite National Park region developed a field medical kit for use by EMT personnel. It can be carried in a single medium sized pack with an internal frame or carried by several members of the unit since the kit is divided into modules. This kit provides a starting point for planning a high-altitude medic's kit.[29]

DIAGNOSTIC MODULE	
Scissors	Two thermometers (clinical and hypothermia)
Blood pressure cuff	Two pair tweezers
Stethoscope	Three airways (adult, child and pedi)
Watch with sweep second hand	One syringe bulb
Penlight	One 50 cc suction syringe and catheter
Hemostat	
SIGNAL & SURVIVAL MODULE	
Flare gun and three flares	Matches
Orange hand-held smoke flare	Pencil and pad
Signal mirror	Yellow plastic tube tent
Whistle	Two space blankets
Compass	Maps
Knife	Toilet paper
INTRAVENOUS MODULE	
Two 1000 cc bags lactated Ringer's	Two 19 gauge butterfly needles
One 500 cc bag 5% dextrose in water	Two 21 gauge butterfly needles
Two Macro solution sets	One roll 13mm tape
One Pedi solution set	Five gauze pads, 5x5mm
Two 20-gauge catheter needles	Ten Band-Aids
Two 18-gauge catheter needles	Ten alcohol swabs
Two 16-gauge catheter needles	Four towelettes
Two 14 gauge catheter needles	Three tourniquets

TRAUMA & DRESSING MODULE	
Two triangular bandages	Six towelettes
Two Kerlix	Ten gauze pads, 10x10mm
Four Kling	Ten Band-Aids
Six Surgipads, 20x19 cm	Two Steri-strips, 13mmx100mm
Three rolls of tape	Two Steri-strips, 6x75mm
One Ace bandage	One pair bandage scissors
One Betadine scrub	Ten ammonia inhalents
Three swabs	One 25X75 cm large trauma dressing
OXYGEN MODULE	
One D tank of oxygen	
Oxygen mask and nasal cannula	
Two oxygen bottles	
DRUG MODULE	
Injectable:	Syringes & needles
Two Meperidine HCL (Demerol) 100 mg	Three 3 cc with needles
	Two 1 cc with needles
One Benadryl 50 mg	Five alcohol swabs
One Xylocaine (Lidocaine) 100 mg	Oral:
Two Narlozone (Narcan) 0.4 mg	15 aspirin tablets
Two Epinephrine (adrenalin) 1:1000 1 mg	10 Seconal tablets, 100 mg
	10 Dexadrine tablets, 5 mg
One Valium 10 mg	10 Codeine tablets
One 25g. Dextrose in 50 cc preloaded syringe	20 Salt tablets
	20 Lomotil tablets
Two bicarbonate preload	Syrup of Ipecac
Topical Neosporin ointment	Activated charcoal
SPLINTS MODULE	
Two full-leg air splints	One wire ladder splint
One arm air splint	One cervical collar
One ankle air splint	One towel for cervical collar
One wrist air splint	Trauma dressing

Thinking lofty thoughts

The United States military is faced with a variety of challenges as it converts from a forward-deployed to an expeditionary force. Part of being an expeditionary force is planning and preparing for contingency missions in various regions. One possible contingency is deployment to a high-altitude region. US Army regular and National Guard units that currently train in the mountains and Alaska are already aware of some of the problems involved in working at altitude and in the cold. They know that such a contingency mission would be difficult, but with proper foresight and preparation, medical personnel can meet the challenge and protect the force.

Notes

1 *The Houghton Mifflin Dictionary of Geography: Places and Peoples of the World*, Boston: Houghton Mifflin Company, 1997, 48 and 450.
2 Tim J. Setnicka, *Wilderness Search and Rescue*, Boston: Appalachian Mountain Club, 1980, 620.
3 Syed Ishfaq Ali, *Fangs of Ice: The Story of Siachen*, Rawalpindi: Pak American Commercial PVT) Ltd., 1991, 15-16 and 110-111.
4 Salman Beg Punjab, "Operations in Glaciated Areas," *Pakistan Army Journal*, Vol. XXXV, No. 1, Spring 1994, 5.
5 Comments of Lt. Keith E. Thompson, USMC Mountain Warfare Training Center, Bridgeport California, in December 2002.
6 Punjab.
7 Setnika, 623.
8 Ali, 16-17 and Punjab, 5-6.
9 Pravin Sawhnee, "Kashmir's Cold War," *Jane's International Defense Review*, December 1997, 59.
10 Punjab, 5 and Thompson.
11 Lester W. Grau, *The Bear Went Over the Mountain: Soviet Combat Tactics in Afghanistan*, London: Frank Cass Publishers, 1999, 193-196.
12 Thompson.
13 Setnicka, 618-619.
14 Boris Gromov, *Ogranichennyy kontingent* (*Limited Contingent*), Moscow: Progress Publishers, 194, 186.
15 Setnicka, 612-618.
16 Thompson.
17 Ibid.
18 Setnika, 609-612.
19 Setnika, 621-623; and V B. Malkin and V. V. Litovchenko, "O profilaktike gornoy bolezni" ("Prevention of Mountain Sickness"), *Voenno-meditsinskiy zhurnal* (*Military-Medical Journal*), September 1982, 38.
20 Setnika, 621-623.
21 Thompson.
22 Setnika, 621-623; Malkin and Litovchenko, 38; and A. M Murzaliev and P. N. Gol'dberg, "Lechebno-evakuatsionnye meropriyatiya pri ostrykh narysheniyakh mozgovogo krovoobrashcheniya v gorakh" ("Treatment and Evacuation Measures for Acute Disruption of Blood Circulation to the Brain in the Mountains]") *Voenno-meditsinskiy zhurnal* (*Military-Medical Journal*), December 1991, 19. Also, see Dennis La Ravia, Uzvaiantha Ganji, Robert Lahasky, and Lynn Pittman-Cooley, "Review of Cerebral Edema: High Altitude Illness," *US Army Medical Department Journal*, January-February 2000.
23 Setnika, 623.
24 Ali, 12.
25 Pravin Sawhnee, "Kashmir's Cold War," *Jane's International Defense Review*, December 1997, 60.
26 Ali, 13.
27 A. V. Dergunov and V. M Didichenko, "O podgotovke vrachey-gornospasateley" ("Preparing Physician Mountain Rescuers"), *Voenno-meditsinskiy zhurnal* (*Military-Medical Journal*), May 1991, 16.
28 Dergunov and Didichenko, 16-17.
29 Setnika, 599-602.

AVIATION

22

Helicopter Landings in the Mountains Lessons From the Soviet-Afghan War

LTC (Ret) Lester W. Grau, CW4 Anthony Reed, and CW3 James W. Grau

Flying helicopters in Afghanistan is a challenge. Its difficulty and danger are usually underestimated by aviators who have not experienced it firsthand. Most airfields and heliports are located from 1,000 to 1,800 meters (3,000 to 6,000 feet) above sea level. The omnipresent abrasive dust eats rotor blades and engines. In the summer, in the south, temperatures reach 45° to 52° Celsius (113° to 126° Fahrenheit). In the north, the summer temperatures reach 40° to 45° Celsius (104° to 113° Fahrenheit). Strong winds are common in the afternoons and evenings, especially in the central and western parts of Afghanistan. These winds reduce visibility and create dust storms. The average mountain range is some 3,000 to 4,000 meters (10,000 to 13,000 feet) high. These flying conditions reduce engine power and lift capacity, lower the flight ceiling, complicate takeoff and landing, and decrease the reliability of the aircraft.[1]

At the peak of the Soviet-Afghan War, the Soviets deployed three helicopter regiments, seven separate helicopter squadrons, and one helicopter detachment to Afghanistan.[2] *The deployed helicopters were the Mi-24 helicopter gunship, the Mi-8MT armed transport helicopter, the Mi-6 heavy transport helicopter, and the Mi-9VZPU airborne command post.*[3] *During the nine-plus years that the Soviets were in Afghanistan, they lost 329 helicopters (127 Mi-24, 174 Mi-8MT and 28 Mi-6). Part of the reason for the loss was pilot fatigue. The sortie rate for pilots reached six to eight flights in a 24-hour period. This worked out to 600-800 flights a year with over 1,000 hours flying time (US aircrews are currently meeting the same amount of hours on deployments). The tour of duty for pilots was two years.*[4] *Pilot fatigue was not the only problem. The enemy, the special flying conditions and the pre-tour training all contributed to the problem. It took time and experience flying in the mountains for bold pilots to have a chance at becoming old pilots.*

Flying in mountains adds additional demands on the pilot. The Russian publication, Army Digest, *recently published the following article about their experiences of flying and landing helicopters in Afghanistan's mountains.*[5] *The authors hope that this information will*

be useful to our own rotary wing community when flying in Afghanistan's mountains or in any other high altitude environment.

Landing a Helicopter in the Mountains

Colonel D. B. Zipir in *Armeisky sbornik*, January 2010

The helicopter's ability to land on small mountain sites makes it an indispensable system for combat and special missions. Training for a mountain landing is one of the basic and yet most complicated forms of training. It demands thorough training and the consideration of a whole range of conditions by the crew. Among these are weather conditions; actual helicopter takeoff weight; the site's suitability for landing and takeoff; the tactical situation; and the enemy's air defense capabilities. At the same time, there must be close coordination among the crew, within the tactical team and also with the forces for whom the flight is being conducted. Statistics emphasize the difficulty of this type of flying: the largest number of serious incidents, accidents and catastrophes occur during the approaches, landings and takeoffs at these sites.

This article is based on the experience of flights and landings at small sites away from an airfield. Its purpose is to inform commanders and flight personnel about the special conditions encountered when landing helicopters in the mountains.

Special features of landing site selection from the air

Evaluate a site's suitability for landing by flying over it at a speed and flight altitude close to the minimum allowed [Effective Translational Lift], since this is optimal for inspecting landing conditions. As a rule, evaluate the overall landing zone characteristics during the first pass. Make a second pass to drop a smoke canister and make

Site Elevation (Meters)	Site Dimensions (Meters)					
	Mi-8	Mi-24	Mi-26	Ka-50	Ka-27	Mi-2
0	—	—	65x80	50x130	50x130	75x200
up to 500	—	50x130	65x250	50x130	50x130	75x240
up to 1,500	55x120	50x175	65x285	50x150	50x150	75x280
up to 2,000	55x165	50x240	65x300	50x180	50x180	75x320
up to 3,000	55x255	50x270	65x315	50x300	50x300	—
up to 3,500	55x300	50x340	65x350	—	—	—
up to 4,000	55x345	—	65x365	—	—	—

Figure 1 Minimum dimensions for a helicopter landing site at different elevations. These values are figured for sites with bordering obstacles of a height of AH≤15 meters, [AH being the equivalent of "Above Touchdown Level."]

a more detailed inspection of the site's surface and obstacles on this and other approaches.

When it is necessary to select a site in a canyon, fly over the canyon at an altitude and speed that allow executing a turning maneuver to the necessary heading for continued flight in the canyon and for a 180° turn for an inspection approach to the site and to drop a smoke canister.

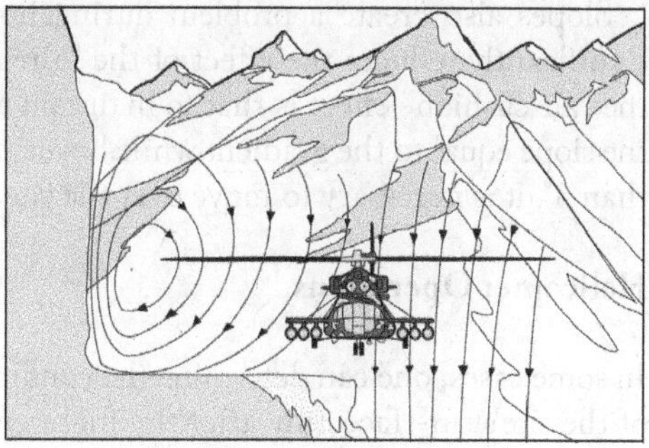

Figure 2.

The minimum dimensions of helicopter landing sites at different elevations are listed in Figure 1. These landing sites are restricted by bordering obstacles of a height of AH≤15 meters.

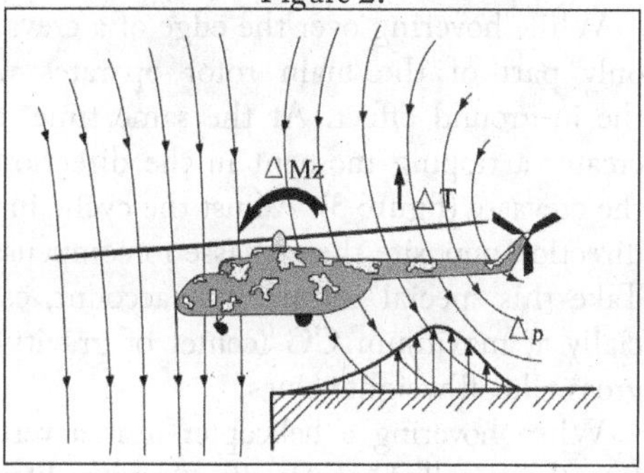

Figure 3 Main rotor flow characteristics while hovering over a crevasse.

Usually the obstacles at the edges of mountain landing sites are natural. They are hillsides, large stones and sometimes trees. It is necessary to evaluate the obstacles both in terms of flying over them with a ΔH>10 meters [altitude correction greater than 10 meters][6] during a landing approach and taking off from them, as well as in terms of their influence on rotor lift while hovering. Remember that, when flying by a hillside, the main rotor axial airflow flow conditions will be similar to a "vortex ring state," and crevasses at the edge of a landing site decrease the effect of the in ground effect "air cushion." Main rotor lift decreases in both cases (figures 2-3). In most cases, one cannot count on the use of the in ground effect cushion at mountain landing sites.

Hovering close to any restricting surface imposes peculiarities on a helicopter's behavior and the piloting technique. If the angle of a surface slope changes, then the influence of this surface from 0° to a slope angle of 45° will have an impact, on the "air cushion effect" — from 45° to 135° as a "vertical face effect" (figure 4), and from 135° to 180° as a "ceiling effect" (figure 5).

The special feature of evaluating slopes is the fact that only from H<5 meters [H (height) is less than 5 meters] can one evaluate a gradient with sufficient precision, and consequently, determine a site's suitability. It might seem readily apparent from H=100 meters, that a site is suitable for landing a squadron of helicopters, yet from H=50 meters it becomes apparent that it will only handle a flight [two helicopters], while at H=3 meters, it is clear that it will be difficult to select a relatively flat place for one helicopter. Be ready for this.

Slopes also create a problem during hovering. First, a gradient greater than 5° significantly reduces the effect of the "aircushion" influence; second, a peculiarity of the "air cushion" effect is that to maintain a hover, one must bank in the direction of the slope equal to the gradient while hovering near the ground. With gradients greater than 3°, it is necessary to move over the site using hover taxiing.

Helicopter Operations

In some cases, one can determine the condition of the site's sur- face only after the flight engineer exits the aircraft.

While hovering over the edge of a crevasse, only part of the main rotor operates with the in-ground effect. At the same time, this creates a tipping moment in the direction of the crevasse (figure 3). Adjust the cyclic in the direction opposite the crevasse to compensate. Take this special feature into account, especially at maximum CG (center of gravity) or gross takeoff weight values.

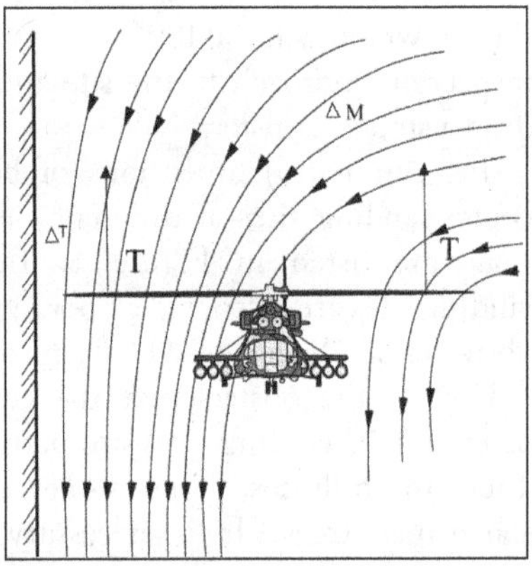

Figure 4.

When hovering a helicopter near a vertical face, the so-called "nozzle effect" occurs. Flow at the face distorts and vertical speed components increase in the plane of the blade's rotation, which leads to a decrease of local angles of attack and aerodynamic loading on the blades passing close to the vertical face. As a result, a moment emerges directed toward the face. The moment's magnitude increases and renders a strong influence from a distance of 0.6-0.7 Ø NV [0.6 to 0.7 diameter of the main rotor blade].[7] To counteract the moment, the pilot must adjust the cyclic in a direction away from the face (figure 4). During certain circumstances (high mountains, site location near a vertical wall, maximum loading, a weak power plant, etc.), there may not be enough power to overcome this moment.

Sometimes a helicopter has to hover under a surface (for example, an overhanging rock ledge, creating the so called "ceiling effect" (figure 5).

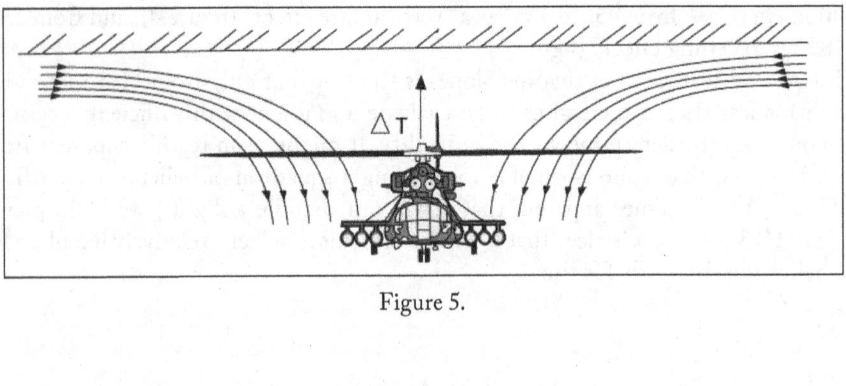

Figure 5.

The ceiling effect deforms the inductive velocity fields in the plane of rotation, due to the characteristic flow. Inductive velocity decreases because of the change of the flow's direction absorbed by the blade. The angles of attack on the blades increase, and, consequently, the main rotor lift grows. The helicopter tends to "adhere" to the overhanging surface.

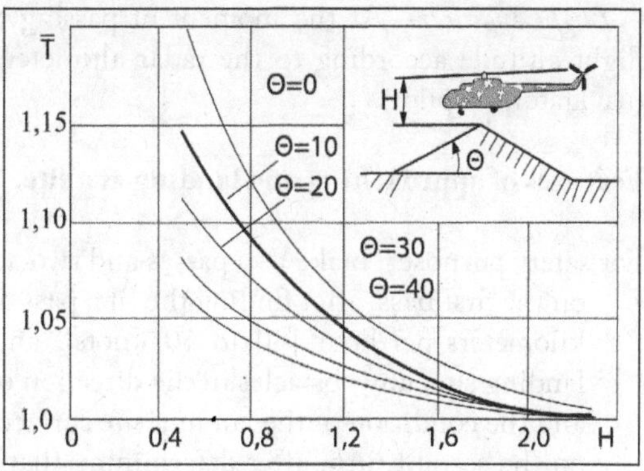

Figure 6.

The landing site's surface should allow landing or hovering over it at a set altitude. The nature of a site's surface influences the conditions of the "air cushion" formation. Gradients close to maximum values significantly reduce the influence of the "air cushion" effect. Rocks measuring more than 15-20 centimeters erode the "air cushion," create problems in selecting a landing site, and require hovering the helicopter at H>1 meter with subsequent deviations (figure 6).

Operator's Manuals for all types of helicopters limit the permissible wind velocity for landings and takeoffs from sites with gradients. It is 5 meters-per-second for takeoff and landing in any direction. At U>5 meters-per-second [Wind velocity is greater than five meters-persecond], takeoff and landing must be made only into the wind.

It is important to evaluate the surface of a landing site for the possibility of formation of dust vortices during hovering (brown out).

Features of crew actions at the landing site selection stage.

Aircraft commander (Pilot in Command)—Conduct the approach to inspect the site and drop a smoke canister in accordance with the heading, taking into account the direction and velocity of the wind and the terrain. At the same time, the estimated lateral wind component should not exceed those limitations established by the Operator's Manual for hovering.

Co-pilot— During flight over the landing site, in accordance with the true altitude reported by the flight engineer and according to readouts of the barometric altimeter, calculate and report to the aircrew commander the "zero" landing site altitude according to the barometric altimeter. Calculation formula:

Ho =Hbar, -Hi , where
Ho — site "zero" altitude;
Hbar —barometric altimeter readouts;
Hi — radar altimeter readouts.

Flight engineer—At the moment of passing the landing site, determine the true flight altitude according to the radar altimeter and report it to the pilot-navigator (navigator, co-pilot).

Features of approaching and landing at a site.

For safety purposes, make two passes and two approaches to the site.

- on the first pass, after finding the site, pass over it with a speed reduced to 80-100 kilometers-per-hour [40 to 50 knots], and determine the dimensions of the landing site, any obstacles in the direction of the expected approach and takeoff, and the condition of the landing site surface;
- on the second turn, after determining that the landing site is suitable, make an approach to drop a smoke canister and conduct a detailed inspection of the site;
- on the third turn, conduct a pass to determine the direction and velocity of the wind according to the smoke trail;
- and on the fourth turn, conduct the landing approach.

Approach parameters for all turns are the same as those described for the first pass. Decrease the flight altitude and speed on the approach glide slope for inspecting and dropping the smoke canister to the minimum values allowed for the landing site conditions.

Drop a smoke canister on each potential landing site, but insure that they are not located close to each other. When using the NSP canister, your time is severely limited for determining the direction and velocity of the wind since it has a short burn time.[8] If you use the PSh-type canister, use black or orange smoke.[9] Poke no fewer than four holes into each PSh canister (if there are less, the canister may not ignite because of insufficient oxygen at those landing sites with an elevation greater than 2,000 meters [6,500 feet]) and loosen the contents of the canisters in the openings that you made.

Select the approach direction, if possible, along ridges and canyons, and over terrain with fewer fractures but with the same lighting.

After passing over the site and determining that it is a potential landing site, make an approach to drop the canister.

Climb to 50-100 meters after dropping the canister, then turn back to the site to determine the wind conditions from the smoke trail. At H=150 meters relative to the site, bring the helicopter into level flight. There is sufficient time while you climb and turn back to the site for the canister to ignite and to form a smoke trail (figure 7). After determining the wind according to the smoke trail, make your landing approach by one of four methods depending on the wind (figure 8). During your next pass, evaluate any obstacles on the approach and during a takeoff into the wind. While making approaches at V=80 kilometers-per-hour [40 knots] to sites with an elevation greater than 2,000 meters, one must be guided by tgp[10] for V=100 kilometers-per-hour [50 knots] because of adjustments to instrument speed.

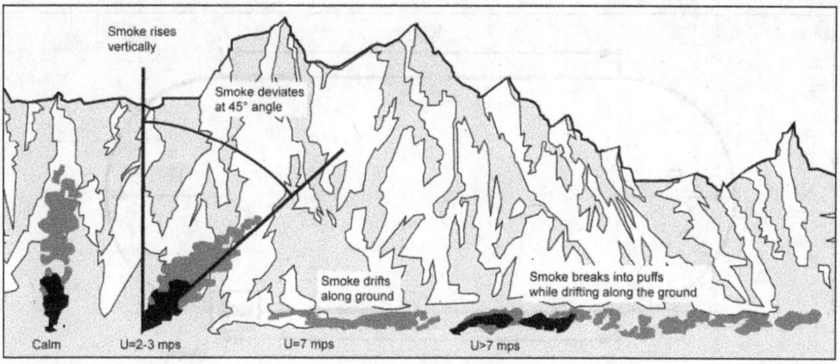

Figure 7.

In mountainous terrain, it is impossible, in the majority of cases, to maintain a stationary hover while viewing the ground just ahead from the left side of the helicopter. This is due to the limited dimensions of the landing sites — and the location of the helicopter's "wheel mounting." This may occur at convex or concave landing sites, and also at sites constrained by obstacles. Therefore, it is necessary to hover and maintain a stationary hover relative to a reference point.

Consider the helicopter's sluggishness during a landing approach the increased pilot response and reaction times required for unforeseen changes in flight parameters. Here, a pilot should expect a longer time in changing aircraft speed after changing his angle of pitch, than when flying at lower altitudes. The greater the flight altitude, the more sluggishly the helicopter performs. Altitude affects a pilot's perceptions and sensations. When flying at H=1,500 meters [4,920 feet] in calm weather, it feels like the approach is being made with a tail wind of U≈3 meters-per-second and at H=2,500 meters [8,200 feet], with a tail wind of U≈5 mps (deduct this accordingly from the speed of a head wind).

A relatively steep glide path allows approaches with less collective pitch applied, but this profile requires larger pitch angle changes to adjust forward speed, and therefore larger engine control inputs. This causes difficulties if the pilot does not have the required skills.

It is necessary to maintain forward airspeed at or below V=50 kilometers-per-hour indicated [27 knots], bearing in mind the increase between your indicated and true air speeds, above V=50 kilometersper-hour — according to the Doppler and drift-angle UMS.[11] Maintain the approach angle by keeping the intended landing site in the cockpit window. Also evaluate the helicopter's position according to the artificial horizon indicator during the third approach. Check main rotor RPM during transition to in ground effect airflow. It is easier during landing when the helicopter's nose is heavy: speed is decreased in the smaller angles of approach, and the field of view of the landing area becomes better.

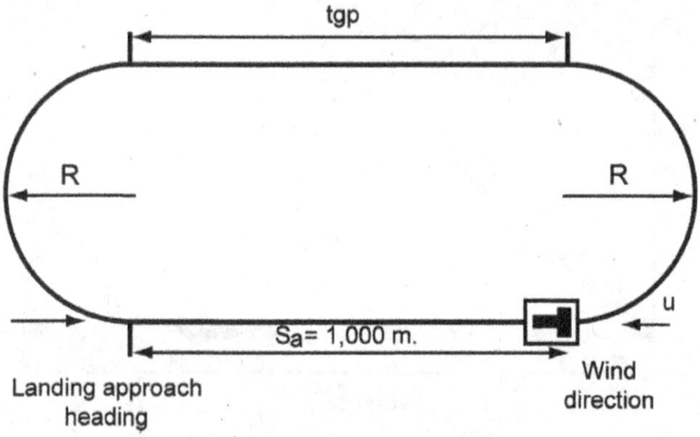

With two 180-degree turns

H=150m.
V= 80(100) kph
V= 22,2(27,8) m/s
Φ=15(10)
$R_{\Phi 15}$=190(295)m.
$R_{\Phi 10}$=285(445)m.
$t_{360\Phi 15}$=54(66)sec
$t_{360\Phi 10}$=80(100)sec

$$tgp = \frac{S - u \times t_{360}}{V+u} + 5sec$$

V, kph		U, m/s	0	3	5	7	10	15
80	Φ=15	tgp, sec	50	38	31	26	19	10
	Φ=10		50	35	27	20	11	0
100	Φ=15		40	31	25	20	14	5
	Φ=10		40	28	20	17	0	0

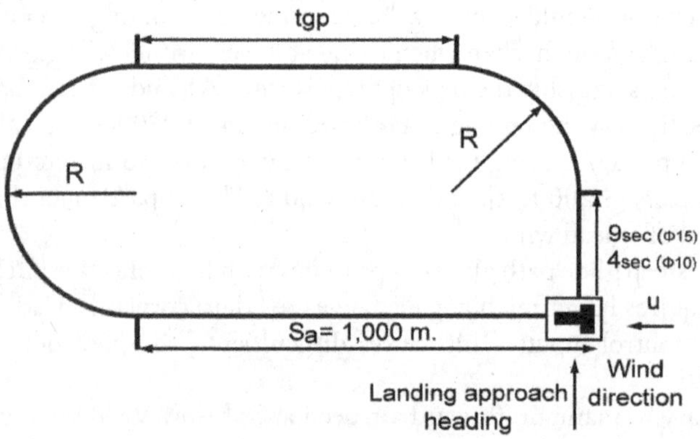

Small Box Pattern

H=150m.
V= 80(100) kph
V= 22,2(27,8) m/s
Φ=15(10)°
$R_{\Phi 15}$=190(295)m.
$R_{\Phi 10}$=285(445)m.
$t_{270\Phi 15}$=41(49)sec
$t_{270\Phi 10}$=60(75)sec

$$tgp = \frac{S - u \times t_{270} - R}{V+u} + 5sec$$

V, kph		U, m/s	0	3	5	7	10	15
80	Φ=15	tgp, sec	41	32	27	23	17	10
	Φ=10		37	26	20	15	9	0
100	Φ=15		30	23	19	15	10	0
	Φ=10		25	16	10	6	0	0

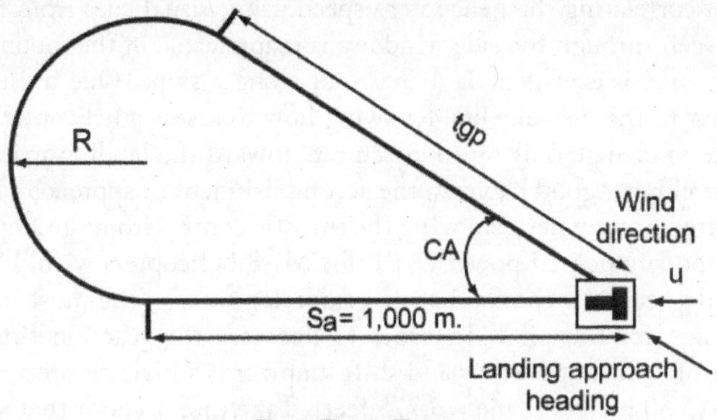

Tear Drop Turn

H=150m.
V= 80(100) kph
V= 22,2(27,8) m/s
Φ=15(10)
$R_{\Phi 15}$=190(295)m.
$R_{\Phi 10}$=285(445)m.
tgPY=2R/S

$$tgp=\frac{S-t_{(180+PY)} \times u}{V+u \times \cos PY}+5sec$$

V, kph		U, m/s	0	3	5	7	10	15
80	Φ=15	CA	21	23	24	26	28	33
		tgp,s	50	41	37	32	27	20
	Φ=10	CA	30	33	37	41	49	63
		tgp,s	50	40	34	29	22	13
100	Φ=15	CA	30	32	33	35	38	44
		tgp,s	40	34	30	26	22	15
	Φ=10	CA	42	45	50	55	63	78
		tgp,s	40	32	27	22	16	7

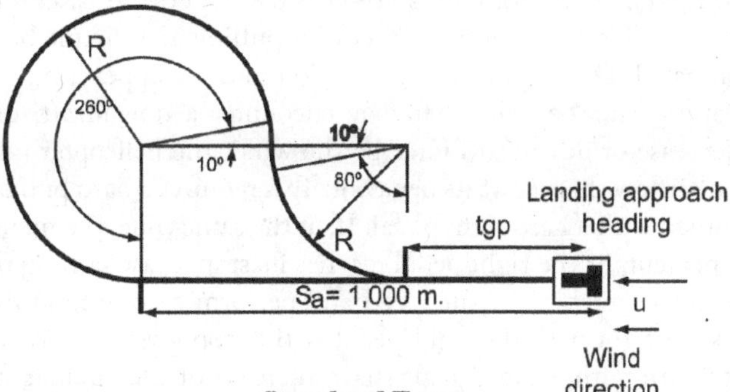

Standard Turn

H=150m.
V= 80(100) kph
V= 22,2(27,8) m/s
Φ=15(10)
$R_{\Phi 15}$=190(295)m.
$R_{\Phi 10}$=285(445)m.
$t_{360\Phi 15}$=54(66)sec
$t_{360\Phi 10}$=80(100)sec

$$tgp=\frac{(1000-2R)-u \times t_{360}}{V+u}+5sec$$

V, kph		U, m/s	0	3	5	7	10	15
80	Φ=15	tgp, sec	33	23	18	13	7	0
	Φ=10		24	12	6	0	0	0
100	Φ=15		20	12	7	0	0	0
	Φ=10		9	0	0	0	0	0

Figure 8 Methods of approach to a site depending on wind direction.

The habit of correlating the helicopter's speed using visual cues from the ground's movement as seen through the side windows is inapplicable in the mountains — the approach, as a rule, is executed via a chasm or along a slope. One must control the speed according to the instruments. Knowing how to assess a helicopter's speed and observe its rate of change only by approach rate toward the landing markers— is an indicator of developing good habits in the accomplishment of approach planning.

Problems often occur when removing the throttle control from the stops. The need to switch on the augmented power (AP) (for Mi-8 helicopters with TVZ-117MT engines) usually occurs at the final stage of deceleration — the most intensive and demanding stage for the pilot. In order to decrease the reaction time required, remove the throttle from the stops and shift it upwards a little before approaching a site with LN>1,760 meters [Line = 5,775 feet]. Take into account that a small shift of the throttle upwards, even without the blinking of the AUGMENTATION ON advisory light, augments the power plant's power. If the pilot forgets afterward to return the throttle control to the stops, then he will have increased fuel consumption. Bear in mind that even the TVZ-117VM and TVZ-117V engines (Mi-24), for the most part, also react with an increase of power at the movement of the throttle upwards, even though power is supposed to be exclusively controlled by the "automatic device."

Know how to use the following feature when necessary. A moveable stop control system is supposed to allow shifting the angle control rod of the tail rotor system through its entire range. However, a malfunction of the "automatic device," an error of the air pressure sensor caused by the "air cushion" while hovering, or the incorrect regulation of directional control inputs are possible. In critical cases such as these, switch off the moveable stop control system by pulling the circuit breaker on the pilot's electrical panel. Doing so locks the position into the upper limit.

During an approach, the helicopter may encounter a downdraft, or, due to an abrupt wind decrease or downward force by the wind, the helicopter may encounter main rotor axial airflow that skews its behavior. This requires a hard pull on the collective pitch combined with a hard right pedal. If, at the same time, the main rotor RPM falls below 92-percent, or the right pedal reaches its stop, cease pulling pitch and use the reserve altitude available to gain speed and perform a go-around for a repeated approach. Banking right with the right pedal at the stop gives a short-term effect of cessation of the turn with a subsequent sharp increase of the angular velocity. This occurs during a decelerating turn as a result of the need to increase collective pitch due to the decrease of lift (Tu) during the bank and to maintain glide trajectory. You should perform a go-around, if possible, by turning away to the left and decreasing collective pitch.

In the other case, if the helicopter gets into an updraft or the winds increase, there is an increase of main rotor lift, the helicopter gains altitude (decreases vertical descent speed), and, to preserve the glide slope, one must vigorously decrease collective pitch. The correct reaction in this case, after passing L=250 meters to the site, is to go around for another approach.

During a landing approach using maximum power or with an inaccurate assessment of the power plant's capabilities, there is a certain "point of no return" on the glide slope. Once this is crossed, the pilot loses the ability to correct the situation by terminating the approach with a go around. One can get into this situation in mountains not only for subjective reasons — errors in the pilot's technique and assessment of available engine power — but also for reasons independent of the pilot. For example, the winds at the final stage of the approach allow sustaining glide slope parameters, but then the winds drop sharply before the helicopter has reached the site ground effect. The helicopter is in a flight profile with no surplus power. The pilot's instinctive reaction is to increase collective pitch, trying to sustain the glide slope. This leads to main rotor droop, an increase of the vertical descent speed or pushing the right pedal to the floor and changing the helicopter's heading. When unable to increase speed or to turn away, "disrupt" the main rotor, creating an angle of pitch that will allow a landing with a forward and vertical speed close to zero. In this case, the relatively steep glide path increases the chance of a successful outcome of the flight.

Maintaining a hover in winter conditions, and also at a high-mountain landing site covered with ice is difficult, even if hovering over the site does not create a snow cloud of powdered snow. The whiteness of the snow, the absence of a level natural horizon line and sunny weather all create difficulty in controlling the spatial position of the helicopter, even with a high-contrast reference point. The presence of several contrasting reference points at the site eases the situation. When needed, an automobile tire or another contrasting object should be thrown onto the snow when passing over the site.

Select the landing location after a moving hover at a minimum altitude (down to one meter). When needed, disembark the flight engineer and "set the wheels" according to his commands.

It is permissible to land and take off with a snow cover thickness of 50 centimeters and 15 centimeters for a running takeoff (10 centimeters for the Mi-24). Taxiing along a snow-covered (dust-covered) site during the absence of horizontal visibility is permitted at Vw<10kph [wind velocity is less than 10 kilometers per hour], turns are permitted while taxiing at Vw<5kph, but turns in place are prohibited.

An AI[12] start is guaranteed only up to H=3,000 meters [9843 feet MSL] because the decreased pressure in the AI air system is lower than that minimally required for start up. Do not switch off the engines after landing above this altitude.

Features of crew actions at landing approach and landing stage

Aircraft commander — Depending on winds, make an approach to drop a smoke canister. Use an appropriate approach method from figure 8. After coming out of the 4th turn, order the flight engineer into the cargo hold to prepare to drop the canister. Glide to the site to Hi=15meters, decreasing air speed to the minimum allowed. After the flight engineer's report that he is ready, at the calculated glide slope points, give the command: "Ignite it" and "Drop." Remember the barometric flight altitude during this pass over the site. After the canister has been dropped, put the helicopter into a climb to an altitude 135 meters [443 feet] above the altitude used to pass over the landing site and increase speed to 80 kilometers per hour [40 knots]. After reaching H=50-100 meters, turn back to the site; evaluate the direction and strength of the winds according to the smoke trail; evaluate obstacles along the specified landing and takeoff heading; make a pass over the site; and make the landing approach using one of the methods depending on the winds.

Co-pilot—After defining the winds according to the smoke trail, set the approach heading on the Doppler meter and report the heading and approach method to the aircrew commander. Turn on the coordinates' meter during a pass over the site.

Flight engineer—Upon receipt of the command, "Prepare to drop the canister," go into the cargo hold, put on the safety belt, hook it to the cable, prepare the canister for lighting, slightly open the cargo hold door and report readiness to the aircrew commander. After receiving the command, light the canister fuse and report to the aircrew commander. After receiving the command for dropping the canister, drop it out, close the door, report to the aircrew commander, remove the safety belt and return to your workstation. Monitor engine operating power and report: "Operating power nominal," "Flight state takeoff," etc. Upon the gas temperatures approaching the maximum value, count backwards: "To maximum temperature 30°..., 20°.., maximum temperature." After the hover, go into the cargo hold on the aircrew commander's order, examine the rear hemisphere in the blisters and look for any obstacles beneath the wheels. Report the results to the aircrew commander. On the aircrew commander's order, disembark from the helicopter, move around to the aircrew commander's field of view, evaluate the underlying surface, and with gestures assist the aircrew commander about the direction of the displacement and the landing location. Always remain in the aircrew commander's field of view throughout the process.

Features of approaching and landing at a site in a canyon

Approaching and landing at a site in a canyon is distinguished from the approach and landing at a site on a flat plain or elevation. The distinctions are connected with the fact that approaches to the site always are made along the length of the canyon. The bottoms of canyons, as a rule, slope and the mountain slopes descend to the bottom of the canyon at different angles. The angle of the approach may be variable and the flight path to the approach may be crooked (figure 9).

After finding a site in a canyon, it is necessary during the initial pass over it to evaluate the flight altitude relative to the site according to the two-pointer altimeter (according to the difference of the radio altimeter and two-pointer altimeter readings). Pull the helicopter into a climb, set V=80 kilometers per hour, "press in tight" to one of the canyon faces (for the Mi-8 and Mi-26, the right wall is preferable). After attaining an altitude that allows a 180° turn according to the width of the canyon (figure 10, bring the helicopter into level flight and make the 180° turn. The width of the canyon is the distance from the helicopter to the canyon wall in which the turn is being made.

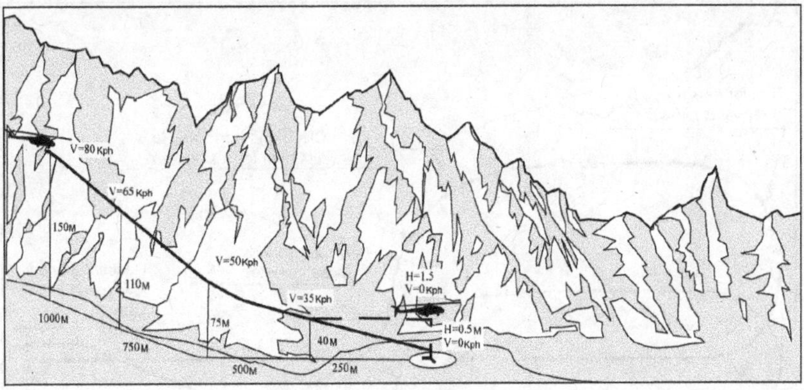

Figure 9.

Φ	W (kph)	Turn R (m)	Turn (m)	Gorge width (m)
20°	80	138	276	380
	100	216	432	540
30°	80	87	174	280
	100	136	272	380
40°	80	60	120	220
	100	94	188	290

Figure 10.

Add 100 meters for error in determining the distance and to compensate for turbulence during the turn.

Calculate your approach so that you can pull out of the fourth turn at a distance that allows your descent in accordance with the parameters of the "low speed" glide slope (depending on flight altitude relative to the landing site). Figure 11 shows the dependence of vertical speed on the speed and altitude relative to the landing site.

Remember that during visual valuation of the distance and elevation of the landing site, flight illusion occurs at a lower altitude and shorter distance relative to the site because of the proximity of the ground's surface and large rocks.

Estimate the flight altitude relative to the site according to the two-pointer altimeter and the distance to the site according to the Doppler indicator. Fly 15 meters over the site, if possible, at the minimum speed for the conditions. Drop the canister and put the helicopter into a climb at V=80 kilometers-per-hour [43 knots] (figure 12).

V at Point of Beginning of Descent (kph)	Vy of Descent (mps) from D=1km		
	From H=100m	From H=150m	From H=200m
120	3.3	5.0	6.7
110	3.0	4.6	6.1
100	2.8	4.2	5.5
90	2.5	3.8	5.0
80	2.2	3.3	4.4
70	1.9	2.9	3.9
60	1.7	2.5	3.3
50	1.4	2.0	2.8
40	1.1	1.7	2.2
30	0.8	1.2	1.7
20	0.6	0.8	1.1
10	0.3	0.4	0.6

Figure 11.

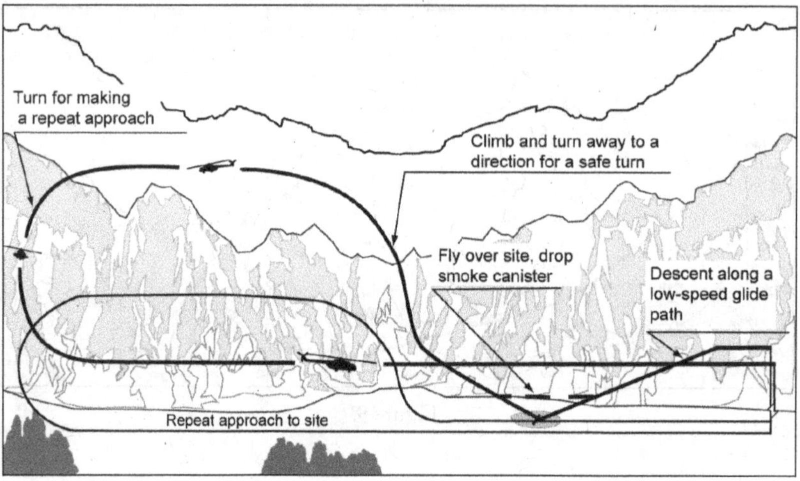

Figure 12.

Pay special attention to maintaining control of the flight during the approach. Flights over inclines or obstacles on mountain slopes are executed with ΔH≥10 m [altitude correction is equal to or greater than 10 meters]. At the same time, the low point of the helicopter, in the majority of cases, is the tail rotor (figure 13).

After a 180° turn, if needed, move to the center of the canyon, deter_ mine the direction and speed of the wind at the ground in accordance with the smoke trail and execute an approach maneuver for landing into the wind.

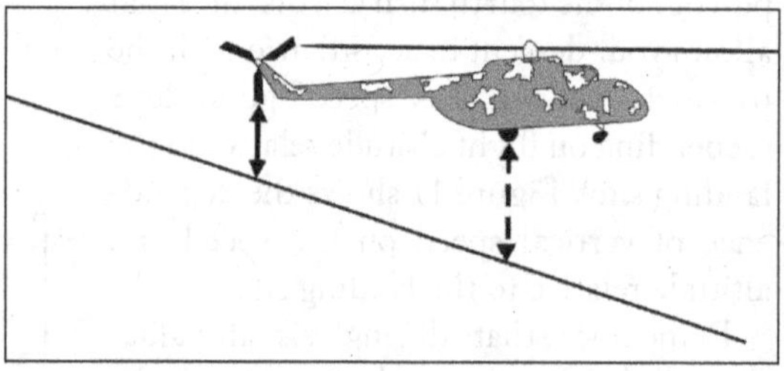

Figure 13

During approach to a "bowl" type canyon landing site, mark out in advance the slope, abeam the hovering location and a readily identifiable reference point. (figure 14). With large angles of closing onto the landing site, do not use a steep glide slope for the approach where V<60kph [velocity is less than 60 kilometers per hour—32 knots] and Vy>3m [vertical speed is greater than three meters per second].

In "box" canyons (those having an approach only from one direction), make the approach for the canister drop at an altitude that allows turning back at V=80 kilometers per hour [43 knots] and maneuver room for a second approach. Evaluate the direction and speed of the wind according to the smoke trail and land. The landing and takeoff are executed with opposite headings (figure 15); therefore, pay special

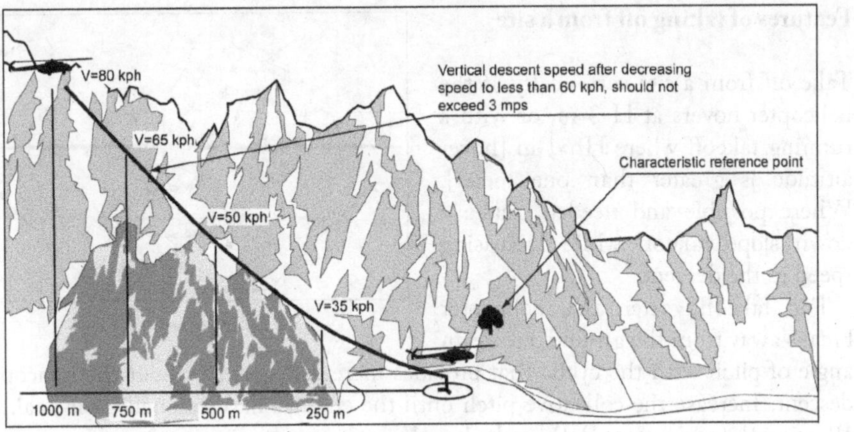

Figure 14.

attention to the direction and speed of the wind. Do not approach and land with U>5m [wind velocity is greater than five meters per second].

Adjust the approach and vertical speed to approximate the parameters of a "low speed" glide slope. Where needed, change of flight direction while following the center of the canyon (figure 14).

The nature of the main rotor's flow in the presence of axial airflow (figure 16) during hovering over a site of limited size at the bottom of a canyon causes a decrease of main rotor lift, which must be con- sidered while evaluating possible landing sites and making landing calculations.

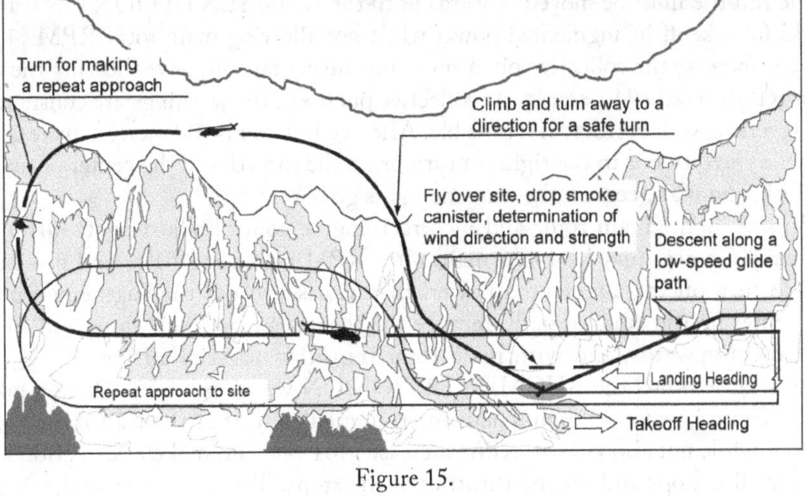

Figure 15.

Features of taking off from a site

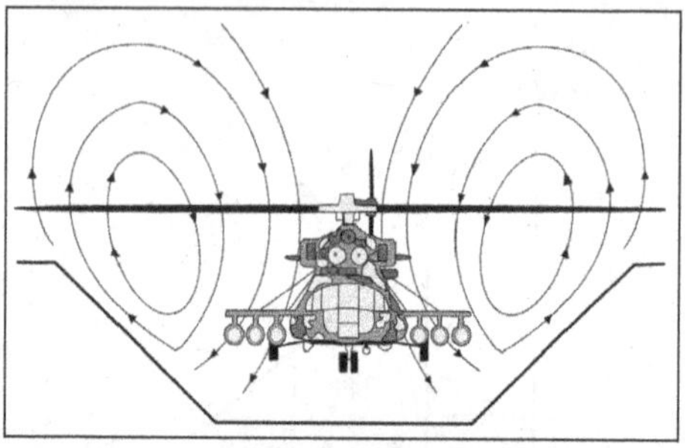

Figure 16.

Take off from a site verti- cally, if the helicopter hovers at H>3 m, or with a running takeoff where Hh>1 m [hover altitude is greater than one meter]. Where possible and needed, make a down slope takeoff while increasing speed in the descent.

For takeoff, after the helicopter breaks away from the ground, create an angle of pitch with the cyclic that provides increased speed without the helicopter's descent. Increase the collective pitch until the engines have reached their peak, not allowing the main rotor RPM to be less than 92 percent.

Use augmented power (emergency power) for landing and takeoff with the Mi-8 helicopter.

Mi-8MT helicopters have TVZ-117MT engines, the maximum power of which is augmented power (AP) that assumes manipulation using the engine controller. Mi-8MTV helicopters have TVZ-117VM engines, the maximum power of which is emergency (EP), and the engines are supposed to reach it automatically.

For an approach using augmented power, after lining up on the landing path, move the throttle upwards until the AUGMENTATION ON warning begins to blink, while not allowing the main rotor RPM to decrease less than 94-percent or increase more than 98-percent. Remember that while augmentation power is on, with collective pitch greater than 8°, there is an automatic correction to the left. This correction must be manually adjusted to the right and checked periodically to insure that it is in the extreme right position.

The throttle must be moved upwards until the AUGMENTATION ON warning blinks for takeoff in augmented power while not allowing main rotor RPM >101%-percent Increase the collective pitch smoothly during takeoff, now allowing the main rotor RPM to fall <94-percent. At collective pitch >7°, the warnings are constantly lit and reversing your correction is possible. After a climb to H=100 meters, make certain that the correction is to the right, in turn bring the throttles to the center position to the lock, and make certain the warning lights go out.

With a "light" main rotor and incorrect engine control adjustment, turning on augmented power may lead to a main rotor RPM increase >101%, and it is necessary to turn off the augmented power. In this case, to use the augmented power, after entering the landing path, remove the throttles from their stops and slightly shift them upwards, and with main rotor transition to axial airflow (V≈35 kph), move it closer until the AUGMENTATION ON warning blinks. During takeoff in augmented power on such a helicopter, immediately after the transition to axial airflow, while not allowing an RPM increase >101-percent, make a correction to the right until it stops and set the throttles at the stops. This all occurs at the stage of the aircrew commander's most intense work in piloting the helicopter. One must be ready for it.

Experience in the use of helicopters with TVZ-117VM engines shows that in the majority of helicopters, moving the throttles upwards also increases engine power, while in the flight manual states that it operates automatically with the augmented power on and there is no need to move the throttles. An increase of main rotor RPM leads to the very same effect on some helicopters with the readjustment. In any case, the pilot must competently use the equipment, know the special features and use all the capabilities for a safe flight.

Evaluation of a helicopter's capabilities in accordance with a power plant's margin of power

This characteristic is most important for the flight's outcome and is difficult for evaluation. It occurs because the existing theoretical methodology does not allow taking into account fully the special features of a specific helicopter and the actual conditions of the approach and landing. In the flight manual, the only correction for wear of the blades during the first compressor stage is turning on the anti-icing system and the dust protection system, which is insufficient in real conditions. Real power plant power and adjustment of automatic fuel controls depend on wear and a number of parameters subject both to change while flying the helicopter and making adjustments. That leads to the fact that the power plant's operational parameters, in a number of cases, do not correspond to the maintenance instructions and are in conflict with the expected results.

Analysis shows that the primary reason for such a situation is inexperience in using the automatic fuel controls. There are no consequences when flying level terrain because of the relatively large power reserve of the power plant. Indeed, it is against regulations to all of the helicopter's power plant and main rotor capabilities in making flights in mountainous terrain and during execution of combat missions.

Often during flights in the mountains, the possibility or impossibility of landing at a site must be evaluated beforehand. The following method is used in this case. If the flight engineer cannot say precisely the maximum temperature of the gases generated by each engine (with a precision to 10° Celsius), then at the site altitude at which the landing will be made, check the augmented power (emergency power) of the engines in flight.

The purpose of the check is to determine the maximum temperature of the gases. The temperature of the gases is the most objective indicator of the reserve power of the engines. Cessation of gas temperature increase always means that the engine has "yielded" all possible power. It is necessary to set V=100-120 kph in level flight for the check, while moving the engine throttles downward, increasing speed in level flight by removing collective pitch until the moment the gas temperature no longer rises or the main rotor RPM has decreased to 92-percent. Remember this maximum temperature obtained from both engines (in practice they almost never are "log book").

During the landing approach, the flight engineer, knowing the maximum temperatures of the gases, indicates to the aircrew commander according to the weaker engine the difference between the actual and maximum values. The aircrew commander, at the same time, must take into consideration that if the gas temperatures have reached maximum, further collective pitch increase will lead to a loss of main rotor RPM or the placement of the right pedal at the stop. This will allow making a decision to land or break-off of the approach in time. One must pay attention to the nature of the change of the power and reserve power required in a hover depending on the elevations of sites and the takeoff weight of the helicopters.

Conclusion

This is an interesting read on mountain approaches and shows that the Russian helicopter pilot still does a lot of "stubby pencil" calculations during flight. US aviators use their Performance Planning Card or PERF page (in glass cockpits) to the same effect. The procedures in the article may also strike the US aviator as overly cautious and time-consuming. The entire article is about approaches and no US aviator is going to routinely make three passes before landing. Aviators can save this precious time during the mission by expending it in extensive training before deploy- ment. The smoke canister methods described in this article, while still used by some of our allies, are not going to be applicable to US Aviators, especially Apache pilots. The co-pilot/gunner is not going to carry smoke canisters in his cockpit, nor is he going to open his crew station door while on approach.

The challenge of high altitude and mountain flying is more than just approaches. One of the problems Colonel Ziper does not discuss is the frequent mountain updrafts and downdrafts that make flying in Afghanistan quite exciting at times. The ceilings on Russian helicopters are usually higher than ours, but the principles of flying remain the same. Even without an enemy, mountain flying is a risky business and the ground unit that we support will not always understand why we need to use certain headings and profiles when coming to their assistance. Power management, properly adapting/reacting to wind shifts, avoiding FOD and an active enemy are all part of mountain flying. So what should aviators do to prepare for a tour in Afghanistan's mountains?

USAACE's The Army Aviator's Handbook for Maneuvering Flight and Power Management *and Chapter 3 of FM 3-04.203* Fundamentals of Flight, *are the bibles for flying in the mountains, (and every aviator deploying to Afghanistan should study them). Afghanistan's flying conditions are not those of Iraq, and a unit should train for mountain flying well before deploying there by simulating conditions during as many flights as possible, both in the aircraft and in the simulators. Even simple traffic pattern flights can be geared towards high altitude training by simulating a lower max torque available. During garrison training, practice maneuvering flight while exceeding the 30°/60° limits in order to become familiar with aerodynamic behaviors of the aircraft such as "mushing." But stay within the 30° / 60° limits during maneuvering flight in Afghanistan. There will seldom be any tactical benefit, and exceeding the 30°/60° limits there can put the crew and aircraft at unnecessary risk. Flying with smooth and slow control movements is optimum, especially in low density altitude environments like Afghanistan.*

Remember that the power margin and bucket speed in a mountain environment will differ from airframe to airframe. It is not in the best interests of CH-47 and UH-60 crews to outfly their AH-64 escort which may have narrower power margins due to gross weights that are significantly increased by their weapon loads. Yet it is common for CH-47s to fly a low-level or contour profile to the base of a mountain and then climb, possibly forgetting that the escort Apache has to start its climb earlier and may be forced to call "heavy metal" in order to catch up and effectively perform its escort mission. Wind direction and velocity are important factors when flying in the mountains, and the winds will vary all the way to the top. Rotary wing aircrews should remember that their AH-64D escort has a wind reading capability which will provide wind heading and velocity at the AH64D's current location. While other aircraft should take advantage of this and any other available winds information before making an approach, it does not substitute for learning good wind and terrain analysis.

There are ten rules of thumb applicable to all helicopter aviators flying in the mountains:[13]

1. *Slow down and perform maneuvers with less intensity for all aspects of the machine. Aggressive left turns and aggressive forward application of the cyclic will increase torque.*
2. *Always leave a way out.*
3. *Know where the winds are.*
4. *Crew coordination is critical. Make sure everyone knows what is going on all the time and split up the duties during the entire flight and during the approach.*
5. *Update performance planning throughout the flight to be aware of power margin, bucket speed and minimum safe single engine airspeed.*
6. *Be aware of hypoxia and take care of your body—get plenty of rest, watch your diet, stay physically fit and avoid tobacco.*
7. *Your senses and perceptions can be unreliable in the mountains— trust your flight instruments.*
8. *Go-arounds are free.*
9. *An aircrew's decision to abort because of adverse meteorological conditions must be accepted by the command.*
10. *Self-removal of aircrew experiencing emotional stress or illness must be encouraged and not criticized.*

Of course, the enemy remains a concern. Afghanistan is one of the few places on the planet where the air defense threat from ground forces may be firing from above the helicopter. During the Soviet-Afghan War, Mujahideen DShK gunners would often emplace multiple machine guns above passes and on flight paths with the hope of downing a helicopter through blade or transmission strike. The Mujahideen also used narrow canyons with some sort of bait in it to draw a single helicopter down away from the support of its wingman. Caves dug into canyon walls contained DShKs which would fill the air with bullets — or a cable might be stretched across the narrow canyon. Potential landing zones were covered with heavy-weapons fire and land mines. Shoulder-fired air defense missiles were often sited at the end of runways. Maintenance test pilots usually fly a predictable pattern in a given area and the Mujahideen were quick to pick up on that.

Flying in Afghanistan is demanding and unforgiving. Like anything else, training, rehearsals, common sense and a bit of caution will assist the bold pilot in becoming an old pilot.

Notes

1 The Russian General Staff (translation, commentary and editing by Lester W. Grau and Michael A. Gress), *The Soviet-Afghan War: How a Superpower Fought and Lost,* Lawrence: University Press of Kansas, 2002, 210-211.
2 Ibid, 318. Viktor Markovskiy, *Zharkoe nebo Afganistana* (*The hot sky of Afghanistan*), Moscow: Tekhnika-Molodezhi, 2000, 101 lists four regiments and ten separate squadrons.
3 Russian General Staff, 210.
4 Ibid, 221.
5 Colonel D. B. Zipir, "Posadka vertoleta na vysokogornye ploshchadki" ("Landing a helicopter in the mountains"), *Armeyskiy sbornik* (*Army Digest*). January 2010, 31-39.
6 It is DH>10 in the original, but is probably ΔH>10. The Cyrillic D [Д] looks similar to Δ.
7 Ø is the diameter, NV is the main rotor blade. In figure 4, ΔT is the lift or thrust differential.
8 The NSP (Nazemnyi signal'nyi patron — ground signal cartridge) is a smoke grenade with a fold-out "leg" base to stabilize the unit.
9 The PSh-M is a smoke pot that comes in black, orange and yellow smoke.
10 The tgp is the flight time required to achieve a horizontal plane before entering the glide path.
11 UMS is probably the "universal multiprocessing system".
12 AI are Aleksandr Ivchenko helicopter engines. The AI-9 is the most popular.
13 The ten rules of thumb are extracted from FM 3-04.203 *Fundamentals of Flight*, FM 1-202 *Environmental Flight,* USAACE *The Army Aviator's Handbook for Maneuvering Flight and Power Management*, AR 40-8 *Temporary Flying Restrictions due to Exogenous Factors.*

Conclusion

Throughout the history of organized militaries, there has been a tension in how best to organize forces for future combat. One school of thought, driven by the budget and the vain hope for simplicity and predictability, have tried to make all combat, combat support and combat service support units the same by branch and function. They would have the same TO&Es, same training, same equipment and, consequently, would perform identically on the battlefield. Recent versions of this school advocate modularity, so that the Defense Department can "plug and play" by attaching and detaching interchangeable units for various contingencies — sort of a national defense Lego© set. Until the current wars in Afghanistan and Iraq, the school's credo was "if units can do maneuver warfare well, they readily can do the "lesser forms of combat" (mountain, counterinsurgency, jungle, urban, riverine) well. The other school sees combat as a complex, hard-to-predict matter that requires well-trained and equipped forces to conduct maneuver warfare plus specially-trained, specially-equipped forces to deal with the lesser forms of combat. In the past, this has led to the development of "elite" forces — paratroopers, rangers, mountaineers, jungle fighters, arctic warriors, special forces, commandos and the like.

These forces are not cheap and, during peacetime, there is often a rift between "big army" and the specialized community. This school's credo is the reason for these specialized forces is that "big army" does not have the time, resources or need to train all soldiers for all contingencies. Of course, elite units are not usually big enough to handle a large contingency over an extended time, so "big army" may then need to train quickly to meet the additional demands of urban, mountain, jungle or whatever combat.

Foreign forces that have experience fighting in the mountains reflect this experience in their training, organization and equipment. Most foreign countries with serious mountains in their country or on their borders have trained mountain units. Their mountain soldiers are trained in far more than basic knots, rappelling, piton placement and individual climbing skills. They are trained in unit mountain maneuver and unit mountain tactical drills; they conduct live fire exercises in the mountains as well as extended exercises in the mountains. Mountain warfare is more than infantry combat and foreign mountain troops train their gunners, sappers, drivers, medics, logisticians and aviators along with their mountain infantry.

The Soviets/Russians have fought two recent wars in the mountains in Afghanistan and in Chechnya. In Afghanistan, most Soviet soldiers were trained in one of seven mountain warfare centers prior to deployment. Their airborne, air assault and special operations (Spetznaz) forces carried the fight to the guerrillas in the mountains. In Chechnya, the initial city fighting went against the Russians, but they rallied, gained control of the urban and built-up areas and eventually moved into the mountains to confront the guerrillas that they had pushed there. It has been a long, brutal campaign, but generally the Russians now control Chechnya and most of the Chechen guerrillas have left the country. The Russians did not control Chechnya until they controlled the mountains. The Russians further fought a quick, if not particularly elegant, campaign against Georgia in the low mountains of Southern Ossetia. This campaign was quick because the Russians understood mountain combat and the need to control choke-points and lines of communication in mountain combat.

India and Pakistan arguably have the best-trained and most-experienced mountain troops, but Chinese Mountain Brigades and Light Infantry Divisions also have experience with insurgents in Tibet and Xinjiang Province. The Columbian Army has a wealth of experience with fighting in heavily-forested and jungle-covered mountains.

This book was written for the mountain training community and for units that will actually fight in the mountains. It is not US doctrine. It is somebody else's idea on how to deal with common problems encountered in the mountains. It is offered for your consideration.

Suggestions for Further Reading

Lt Col Syed Ishfaq Ali, *Fangs of Ice (Story of Siachen)*, Rawalpindi: Pak American Commercial (Pvt) Ltd, 1991. A Pakistani perspective on the fighting Indo-Pakistani fighting in Kashmir over the Siachin glacier — the highest battlefield in the world. Plenty of tactical detail.

Christopher Duffy and Bill Younghusband, *Eagles Over the Alps: Suvorov in Italy and Switzerland, 1799*, Emperor's Press, 1999. Russian mountain campaigning in the Alps by one of Britain's best historians.

A. B. Feuer and Bob Dole, *Packs On: Memoirs of the 10th Mountain Division in World War II*, Mechanicsburg: Stackpole books, 2006. A good history of the training and deployment of the US Army's only mountain division during World War II.

D. M. Giangreco, *The Soldier from Independence: A Military Biography of Harry Truman*, Minneapolis, Zenith Press, 2009. Chapters 12-16 provide an excellent account of a battery fight in the Vosges Mountains.

Lester W. Grau, *The Bear Went Over the Mountain: Soviet Combat Tactics in Afghanistan*, originally published by NDU Press in 1996, now published by Mentor Enterprises of Huntsville, Alabama. Translation and commentary of Soviet tactical experience of Soviet veterans that was compiled by the Frunze Military Academy.

Headquarters, Department of the Army, FM 31-27, *Pack Animals in Support of Special Operations Forces*, USGOVPO, 2000. An excellent guide to the forgotten, but still necessary, skills involved with animal logistics transport in difficult terrain.

Martin Helprin, *A Soldier of the Great War*, New York: Harcourt, Brace and Company, 1991. Novel about an Italian soldier fighting in the mountains during World War I. Well researched and a good read.

Major General D. K. Palit, *War in High Himalaya: The Indian Army in Crisis, 1962*, New Delhi: Lancer International, 1991. Best single English-language book on the 1962 Sino-Indian War in the Himalayan Mountains.

Ali Ahmed Jalali and Lester W. Grau, T*he Other Side of the Mountain: Mujahideen Tactics in the Soviet-Afghan War*, originally published with color maps by Quantico: USMC Special Study, 1998, and now by Mentor Enterprises of Huntsville, Alabama. Like the Marine Corps edition, Mentor's also includes color maps which are integral to understanding the tactics, and the reader is cautioned against *Afghan Guerrilla War* and other unauthorized printings which have poor quality black and white maps. The authors interviewed more than 100

Mujahideen commanders to record their tactical experience in their war with the Soviets.

Andrew M. Roe, *Waging War in Waziristan: The British Struggle in the Land of Bin Laden, 1849-1947*, Lawrence: University Press of Kansas, 2010. Good study of the British struggle in the Northwest Frontier Province and Waziristan bordering Afghanistan.

Russian General Staff (translation and commentary by Lester W. Grau and Michael Gress), *The Soviet-Afghan War: How a Superpower Fought and Lost*, Lawrence: University Press of Kansas, 2001. An operational overview of the Soviet mountain war in Afghanistan and the role of all the branches.

General Sir Andrew Skeen (commentary by Lester W. Grau and Robert H. Baer), *Passing It On: Fighting the Pushtun on Afghanistan's Frontier*, Fort Leavenworth and Fort Benning: Foreign Military Studies Office and Maneuver Center of Excellence, 2011. Reprint of an excellent 1932 guide for British lieutenants and captains fighting on the border with Afghanistan. The terrain and enemy have not changed.

Dodge Billingsley and Lester Grau, *Fangs of the Lone Wolf: Chechen Tactics in the Russian Chechen Wars 1994-2009*, Ft. Leavenworth and Quantico: Foreign Military Studies Office and United States Marine Corps, 2012. Chechen tactical experience based on interviews with Chechen combatants who fought in mountain, urban and open terrain. The book has color maps for clarity.

Index

acclimatization, 19, 20, 27, 31, 37, 39, 211, 212, 214
avalanches, 18, 31, 34, 99, 100, 101, 102, 173, 179

belay, 68, 101, 102, 143, 149, 150, 151, 152, 154, 156, 157, 158, 159, 160, 162, 163, 164, 167, 168, 172, 186, 187, 188, 190, 19,

cerebral edema, 19, 39, 210, 213, 214, 218
climbing injuries, 213
communications, ix, x, xii, 21, 25, 34, 40, 58, 62, 64, 65, 66, 79, 99, 199, 201, 202, 203, 204, 205, 206

defensive actions in the mountains n/a
diesel engines, 19, 24, 175, 184, 204, 210

fire support, xii, 23, 27, 34, 40, 54, 86, 90, 110, 114, 117
flanking detachment, vii, 60, 61, 63, 64, 65, 66
footwear, 58, 168
frostbite, 19, 22, 31, 34, 39, 41, 210, 212

GPS, 65, 66, 105
grassy slopes, 104, 143, 144, 145, 146, 149, 156, 158

helicopters, 17, 19, 20, 23, 24, 25, 35, 36, 40, 53, 83, 84, 85, 89, 111, 116, 175, 184, 195, 202, 205, 206, 210, 221, 222, 223, 225, 230, 236, 237, 238
hypothermia, 19, 210, 213

infantryman's load, 65

Jumar ascenders, 57

leadership, 27, 33, 37, 39, 75, 79, 80, 85, 89, 91, 92, 115, 139, 141
logistics, viii, x, xii, 20, 22, 23, 24, 25, 26, 27, 28, 35, 36, 40, 47, 48, 59, 75, 76, 89, 110, 111, 175, 177, 185, 186, 188, 189, 193, 195, 197, 198, 205, 243

maneuver, xii, 22, 34, 56, 81, 110, 113, 114, 119, 133, 188, 194, 196, 197, 198, 223, 234, 241, 244
medical evacuation, 27
mountain operations, 30, 35, 41, 118, 140, 174, 189
mountain sickness, 57, 213, 214
mules, 20, 24, 25, 26, 35, 36, 41, 59

offensive actions, 22, 33, 93

pace of movement, 57
physical fitness, 31, 33, 37, 173, 174
physiological changes, 18, 210
porters, 20, 24, 26, 27, 35, 41, 175
pressurized sleeping bags, 31
psychological changes, N/A
pulmonary edema, 214

reconnaissance, vii, 33, 35, 47, 48, 50, 61, 84, 94, 97, 99, 100 101, 103, 105, 138, 177, 178, 181, 182, 184, 185, 186, 197
river crossing, 48, 164, 165, 166, 172, 198
rock falls, 101, 102, 179, 196
rope work, N/A

shelter, viii, 19, 52, 75, 86, 135, 137, 138, 210
small-arms fire, 25, 57, 59
solar radiation, 100

talus, 104, 133, 146, 147, 172
thunderstorms, 18, 99, 100
training, xii, xiii, 20, 25, 26, 27, 30, 31, 35, 36, 37, 38, 39, 40, 41, 42, 50, 56, 57, 58, 59, 64, 68, 70, 71, 76, 103, 104, 110, 111, 118, 119, 125, 132, 133, 139, 140, 141, 143, 145, 149, 150, 151, 152, 153, 154, 155, 156, 157, 158, 160, 161, 163, 164, 165, 166, 167, 168, 169, 170, 171, 172, 173, 174, 179, 188, 189, 193, 206, 211, 215, 216, 218, 221, 222, 238, 240, 241, 242, 243
trash, 57

water transport, viii, 175, 184, 189, 190, 191, 192, 193, 194, 195, 196, 197, 198